SUSTAINABLE RETROFITTING OF COMMERCIAL BUILDINGS

T0179150

Whilst sustainability is already an important driver in the new building sector, this book explores how those involved in refurbishment of commercial building are moving this agenda forward. It includes chapters by developers, surveyors, cost consultants, architects, building physicists and other players, on the role they can each play in enabling refurbishment to be commercially, environmentally and socially sustainable. Case studies from northern climates show real examples of different building types, ages and uses and will demonstrate what action has been taken to create more sustainable buildings.

The chapters raise and discuss all the relevant issues that need to be considered in retrofitting decision-making. Changing standards, planning, process management, financing, technical issues, site organisation, commissioning and subsequent building management are all considered. *Sustainable Retrofitting of Commercial Buildings* demonstrates that buildings can be made comfortable to occupy, easy to manage and low in energy demand and environmental impact.

Simon Burton has worked in the field of energy conservation in buildings and urban areas for more than twenty years. He was a Director of ECD Energy and Environment and subsequently a Regional Director with AECOM in London. He has been responsible for several UK government research projects.

SUSTAINABLE RETROFITTING OF COMMERCIAL BUILDINGS

COOL CLIMATES

EDITED BY SIMON BURTON

Routledge
Taylor & Francis Group

LONDON AND NEW YORK

earthscan
from Routledge

First published 2015
by Routledge
2 Park Square, Milton Park, Abingdon, Oxon OX14 4RN

and by Routledge
52 Vanderbilt Avenue, New York, NY 10017

First issued in paperback 2020

Routledge is an imprint of the Taylor & Francis Group, an informa business

British Library Cataloguing-in-Publication Data
A catalogue record for this book is available from the British Library

Library of Congress Cataloging-in-Publication Data
Sustainable retrofitting of commercial buildings. Cool climates / edited by Simon Burton.
 pages cm
 Includes bibliographical references and index.
 1. Commercial buildings—Remodeling. 2. Sustainable buildings—Design and construction. 3. Sustainable construction. 4. Commercial buildings—Energy conservation. 5. Buildings—Energy conservation.
 I. Burton, Simon, 1945–
 TH4311.S87 2014
 690'.24—dc23   2014005759

ISBN 13: 978-0-367-57604-2 (pbk)
ISBN 13: 978-0-415-83424-7 (hbk)

Typeset in Univers
by Keystroke, Station Road, Codsall, Wolverhampton

CONTENTS

LIST OF FIGURES

LIST OF TABLES

NOTES ON CONTRIBUTORS

Simon Burton, BSc, Editor

Simon Burton originally trained as an engineer and urban planner. He subsequently worked in the field of energy conservation in buildings and urban areas for more than twenty-five years. He was a Director of ECD Energy and Environment and subsequently a Regional Director with the Sustainable Development Group at AECOM in London, previous Faber Maunsell. For several years he lived in Brussels and carried out many European Commission supported projects on energy efficiency and renewable energy sources in both new and existing buildings. Low-carbon refurbishment of buildings has been a focus area in the last ten years, leading to editing of 'Energy Efficient Office Refurbishment' in 2001 and writing *The Handbook of Sustainable Refurbishment: Housing*, published by Earthscan/Routledge in 2011.

John Davies, Nat.Dip BSc (Hons) MSc

John Davies is the Head of Sustainability at Derwent London plc and is responsible for creating and leading the company wide sustainability agenda. John is a highly experienced sustainability management professional, with over fifteen years in industry. He is recognised as an expert in several sectors, in particular commercial property, and has developed and led the creation of many industry leading tools and initiatives. He writes extensively in the sustainability press and sits on many industry panels and committees.

Prior to joining Derwent London, John was Head of Sustainability at Davis Langdon where he was responsible for developing and delivering its range of sustainability services and forming strong relationships across its key client base. Before joining Davis Langdon, John worked in the client domain as a sustainability advisor on a range of major projects and organisations, most notably at BAA, where he lead the sustainability agenda within the T5 design phase and the £10bn Capital Projects function as their Capital Projects Sustainability Manager.

Ursula Hartenberger

Having worked on environmental and corporate responsibility issues for a number of global organisations operating in a variety of sectors, Ursula Hartenberger joined RICS in 2006 as Head of EU Policy and Public Affairs, leading the organisation's strategy on energy efficiency, sustainable construction and urban development. In 2009, she took on the role of RICS Global Head of Sustainability and is responsible for coordinating the organisation's strategic activities with regard to capacity building, communication, research and global engagement with decision-makers and sectoral partners. She is member of a series of international sustainable development platforms and stakeholder groups and has been closely involved in RICS publications and initiatives regarding the value implications of sustainability in the built environment and associated investment decision-making. Holding a Masters Degree in Art Market Valuation, Ursula writes for external publications and academic journals and is a regular speaker at international conferences.

Bill Gething, MA (Cantab) Dipl Arch RIBA
Bill Gething is Professor of Architecture at the University of the West of England, an architect and sustainability consultant, having been a long-standing partner of the architectural and urban design practice Feilden Clegg Bradley Studios. He is a Visiting Professor at the University of Bath and an external examiner at the Architectural Association and the Bartlett School of Architecture, UCL.

He was the RIBA President's Sustainability Advisor from 2003 to 2009 and was lead author of the Green Overlay to the RIBA Plan of Work in 2011. He wrote a briefing report in 2010 setting out the agenda for climate adaptation in the built environment to support design teams involved with UK Technology Strategy Board's Design for Future Climate programme and has drawn out lessons learnt in the first tranche of funded projects in his book *Design for Climate Change*.

Nigel Addy, BSc (Hons) MRICS
Nigel is a Chartered Quantity Surveyor and is a Director at AECOM. Nigel has over twenty-eight years of experience in the construction industry covering a broad range of building projects and types.

Nigel is based in London as part of the Commercial Office team and specialises in refurbishment projects. He was on the Steering Group with CIRIA for their publication *Good Practice Guidance for the Refurbishing of Occupied Buildings*. He is also the author of the 'Cost Model for Office Refurbishments' published in *Building* magazine. He has been involved in many refurbishments projects with established leading property developers in London such as Derwent London and Land Securities, and buildings such as The Johnson Building, Page Street, 80 Charlotte Street and 20 Eastbourne Terrace.

He is experienced in understanding the cost drivers for refurbishment in order to provide clients with the right strategic advice for a successful project.

Dan Staniaszek, MA (Oxon) MSc CEng MEI
Dan heads up Sustainability Consulting Ltd, offering freelance consultancy services in the sustainable energy space. He is currently working with the Brussels-based Building Performance Institute Europe, where his responsibilities have included heading up BPIE's data management, renovation and financing portfolios, as well as developing BPIE's modelling and scenario

analysis capabilities. His current role builds on a broad experience base spanning twenty-seven years, including fifteen in senior positions. In various advisory roles in the UK public/non-profit sector at national, regional and local level, he has influenced the design of EU Directives and UK energy policy, including the Renewables Obligation and the Energy Efficiency Obligation, and directed a wide variety of initiatives, including evaluation, certification and knowledge management services. He has also worked with numerous UK and international consultancies.

Lizi Cushen, M Arca ARB

Having studied at the Royal College of Art in London, and had previous experience working for acclaimed UK Passivhaus practitioner, Justin Bere, Lizi joined Allford Hall Monaghan Morris in 2010 as a Part II Architectural Assistant. She qualified as an Architect from Cambridge University in 2012 and has a broad range of experience on residential, commercial and public realm schemes including Green Tea, London. Lizi was also Project Architect for AHMM's BREEAM Outstanding offices, completed in 2013.

She takes an active role in AHMM's sustainability working group which has successfully promoted low energy design across the practice and is working to develop the practice's reference library.

Nick Baker

Nick Baker qualified in physics and after a brief period working in medical physics, he moved to building science as a teacher, researcher and consultant. He has recently retired from University of Cambridge Department of Architecture, where he was involved with several EU-funded research projects, mainly in the fields of building energy, daylight, natural ventilation and comfort, on which he has published many papers. During this time he has written several books, including *The Handbook of Sustainable Refurbishment*, and contributed to others on comfort and sustainability.

David Richards, BSc (Hons) CEng

David Richards is a Director of building design in Arup. Originally a mechanical engineer he has developed an understanding of integrated design founded on a rich and varied project experience in the UK, America and Middle East. Since 2012 he has been leading the facades team in London and the UK. Dave's particular skills are the strategic planning of the building form, facade performance and passive low energy design. He has a particular focus on the energy performance of the building envelope and the interaction with the mechanical and passive systems of a building. He has led integrated engineering teams on a variety of projects including headquarter office buildings, commercial developments, cultural facilities, university campus buildings and airports.

Dave has a strong interest in the subject of learning and has been a visiting tutor of environmental design at the Architectural Association, The Bartlett, MIT, Columbia University and the University of Pennsylvania.

Ljubomir Jankovic, BSc PhD FIAP CEng MCIBSE

Ljubomir Jankovic has worked as an academic, researcher and practitioner on instrumental monitoring, dynamic simulation, and environmental design of buildings for almost three decades. He studied for his undergraduate degree at the University of Belgrade and his PhD at the University of Birmingham. He is a Chartered Engineer, a Member of CIBSE, and a Fellow of the Institution of Analysts and Programmers. His book *Designing Zero Carbon Buildings Using Dynamic Simulation Methods* was published by Routledge in 2012. He was conferred as Professor of Zero Carbon Design by Birmingham City University in 2013.

Paul Appleby, BSc (Hons) CEng FCIBSE FRSA

Paul Appleby advises design and master-planning teams on the integrated sustainable design of buildings and communities. He has worked in the construction industry as a consultant, lecturer and researcher for forty-five years, working on award winning projects with some of the world's leading architects and developers. As well as writing some seventy publications, including key guidance published by CIBSE and others, his book *Integrated Sustainable Design of Buildings* has appeared in a list of the Cambridge University's 'Top 40 Sustainability Books of 2010'. His follow up, *Sustainable Retrofit and Facilities Management* was published in January 2013. He is a Built Environment Expert for the Commission for Architecture and the Built Environment (Cabe), involved in Design Reviews for major projects seeking planning approval, and is actively involved with the UK Green Building Council, sitting on its Policy Committee and Retrofit Incentives Task Group.

PREFACE AND ACKNOWLEDGEMENTS

This book is the companion volume to one with the same title for 'Warm Climates' published by Earthscan/Routledge in 2013, edited by Professor Richard Hyde from Sydney University. Many of the buildings issues are similar for both cold and warm climates, and with climate change this is likely to become more so in the future. For this and other reasons, the approach taken in this book is to focus on the buildings industry and examine how the various players are thinking about and responding to the challenge of sustainable retrofitting of our existing non-domestic buildings, be they historic, outdated, unmanageable, energy guzzlers or simply ready to be refurbished.

The authors of the various chapters are experts in their field, drawn from leading companies, universities and other organisations. The content of the chapters is therefore the view of the individual authors and although this inevitably leads to some overlap and differences of opinion, and many stylistic differences, the whole book has been designed and edited to give a comprehensive picture of the subject, from different perspectives. As the authors are mostly the partners, or typically represent the partners, necessary for major refurbishment projects, we can understand how the different approaches and emphases need to be brought together for a successful product.

As well as acknowledging the contributions of all the chapter authors, it is important to remember all those contributing information for the case studies included in Chapter 15. Many people are involved before a good case study can be presented and these case studies are short summaries only of many years of work, leading to insights into the real world of commercial refurbishment.

My thanks therefore go to all the chapter authors, those providing information and text for the case studies, all the organisations providing back up and permission to use their buildings as case studies, and the photographers who in all cases have given their photographs for free. Several other individuals have been most helpful, offering suggestions, information, comment and support and I would particularly like to thank Nic Crawley, Roderic Bunn and Lionel Delorme in this context. Additionally there are others without whose support this book would never have been produced, at Routledge, Nicki Dennis for the invitation and support and Alice Aldous for constant organisational back up, and my partner Daphne Davies for unfailing encouragement and enthusiasm for the project.

INTRODUCTION
Simon Burton

Most of the world's scientific community believe that the apparent climate change is manmade, being largely caused by emissions from the use of fossil fuels, and we know that around 40 per cent of this energy is used in buildings. Reducing energy use in our building stock is thus a major concern, with the associated need to make the buildings resilient to the climatic changes that are already apparent and inevitably will become more severe in the future.

Energy is used in the materials and process of construction and in all aspects of using and managing the building for as long as it is occupied. The latter includes energy used for the building itself – ventilating, lighting, heating, etc. – the electrical appliances and equipment in and around the building, the water used, transport of people to and from the building, and so on. This book focuses on the decision-making and process of retrofitting existing commercial buildings to use less energy in all these areas and how this can be achieved at the same time as enhancing the other aspects of sustainability, related wider environmental and social issues.

RETROFITTING IS MORE SUSTAINABLE THAN DEMOLITION, BUT CAN WE DO MORE?

Why are we interested in sustainable retrofitting of non-domestic buildings? We know that new buildings can be quite easily built to be environmentally friendly, and legislation inexorably moves us in this direction. But in most countries we have a large stock of offices and other non-housing buildings that are certainly not efficient to run nor necessarily comfortable to work in, so their future must be to either demolish them or refurbish them. There are at least three valid reasons for refurbishment rather than demolition: the building may be an important historic building; it may be capable of refurbishment at lower cost than demolition and new build; or it may be considered that the environmental impact is less if it is refurbished rather than demolished.

Many people believe that the argument for making the refurbishment more sustainable rather than simply complying with minimum legal standards is hard to refute. Social pressure, company image, government pressure, future proofing and helping to combat global warming are among the valid and usable reasons. Most concerned individuals want to contribute to a more sustainable environment.

But there are different pressures across sectors: the commercial world has different drivers and ways of operating compared with the public sector and again the owner-occupier may well see their buildings in yet a different way. The public sector is likely to be under pressure to lead the way in demonstrating sustainable buildings, and owner-occupiers may recognise more and longer term benefits from sustainable refurbishment of their premises, particularly operational aspects (see Chapter 14 'Soft Landings'). However, that said, the boundaries between sectors may be blurring as developers take an increasing role in providing buildings for the public sector and more flexibility is required related to renting and subletting to accommodate fluctuations in the demand for space.

The term 'commercial' can be used in two ways: it can mean the ownership and use of the buildings – offices, factories, retail facilities – as opposed to the public ownership or use of buildings for housing or educational purposes, and also the financial viability of the retrofit. This book covers both meanings but this is a focus rather than a strict interpretation, as the public and private sectors can learn much from one another, and what is expensive today may become commercially viable tomorrow. To be effective, sustainability must bring together environmental, financial and social aspects and thus live in the real world of owners, financiers, designers, contractors, building managers and occupants. All these players need to be motivated to act sustainably; they need the knowledge and tools and they need to support each other if projects are going to be successful.

The authors of the different chapters of this book give their perspectives on how their specialism can contribute to sustainable retrofitting projects, and taken together they cover all aspects from finance to comfort, and stages from project conception to occupation and ongoing use. Although some of the authors are writing from a UK perspective or from experience in UK, the issues and arguments are felt to be equally applicable generally to northern and temperate climates. Authors' contributions have been left in their own particular style and there are inevitably overlaps and differences of opinion. Sustainable refurbishment is a developing field and apart from conflicts between different disciplines, there are also different experiences and different solutions that are still being developed. The important thing that is demonstrated in the chapters and case studies is that across the board there is research and debate leading to action and progress towards retrofitting our commercial building stock for a more sustainable future.

SUSTAINABLE RETROFITTING FROM THE PERSPECTIVE OF THE PLAYERS

Chapter 1 gives a developer's view of sustainable refurbishment of older buildings and on current and future requirements for commercial buildings. The market is always changing and space providers need to be ahead of the game, looking at business requirements, employers' and employees' demands and technological developments. Can existing buildings be refurbished to meet these new demands whilst satisfying sustainability criteria? Developers are inevitably very aware of financial aspects and the social aspects as they relate to users, but environmental sustainability aspects are being increasingly enforced by regulations and consumer demand. Existing buildings may be very suitable for refurbishment or, with a different approach to traditional standards, may provide a financially viable and environmentally successful result when retrofitted. Equally, a change of use of old buildings may be the way to bring them back into service without demolition. Refurbishing old buildings can provide very sustainable solutions with adequate professional input, and lessons from older buildings can even be useful in planning new buildings.

Property valuers play an important role in advising their clients regarding the potential benefits of investments in sustainable refurbishment and how sustainability impacts a commercial building's asset value, as described in Chapter 2. Valuers understand the need to incorporate sustainability into their valuations but there are difficulties including lack of actual data and the lack of skill of valuers in this area, and also important is how to make the business case to lenders. Putting a value on risk factors, possible future insurability and understanding why renters want to rent sustainable buildings are all challenges for the valuation profession.

The architectural profession, often as the project managers as well as designers, is always a major player in driving a sustainable refurbishment. Whilst designing to reduce the emissions that drive climate change, the architect must also focus on adaptation, designing maybe in a different way to take proper account of the changes underway and make the building resilient to climate change. Chapter 3 gives an overview of the role and necessary understanding of the architect designer to combat likely climate change and ensure the building will perform adequately for its design lifetime. Information on predicted temperatures is available from various sources but the designer needs to understand the issues relating to each project and steer a route through to a reasonable risk assessment result. Overheating and comfort are the main developing areas but designing for changes in construction and materials performance, and the dual issues of potential flooding and water usage, also need careful attention.

The cost consultant or quantity surveyor is always going to be central in decision-making on whether to refurbish a building or to demolish it, and what type and level of refurbishment is going to provide the best return. Chapter 4 looks at refurbishment from the perspective of the cost consultant and covers all aspects that can come into refurbishment decision-making.

There are always some positive reasons for renovation but not every building is suitable and careful survey is needed. The choice of what level of refurbishment, repair, remodel or renew, will depend on a range of factors, both physical and financial, and should include analysis of what and how additional value can be added as part of the refurbishment. Evaluation of the cost and benefits of a more sustainable refurbishment is another issue. Also, in purely financial terms the possibility of tax allowances for some parts of sustainable refurbishment may be worth researching and incorporating.

There is a large stock of non-domestic buildings across Europe and other northern climates and many of these will need renovating or replacing over the years. The energy demand of these buildings tends to be much higher than new buildings due to increasing regulations with which they must comply. Chapter 5 describes the results of survey work carried out to assess the energy performance of this existing building stock in Europe and looks into what can be done to reduce this. Whilst the floor area of non-domestic buildings is only 25 per cent of the area of the housing stock, the energy consumption per square metre is 40 per cent greater and has increased over the last decade by 32 per cent, particularly due to increase in the service sector. Regulations to reduce energy consumption in buildings apply not only to the new build sector, and the EU is advancing requirements when commercial buildings are refurbished, providing a joint push to achieve the EU's overall energy reduction targets. A survey of construction executives demonstrated considerable activity in the retrofit market, more so in the EU than the USA and China but not much with the specific objective of energy use reduction, though the future value of this was widely appreciated. One EU renovation demonstration project including 300 buildings showed what could be done to improve energy performance, with energy savings of between 26 per cent and 89 per cent recorded.

Architects, being at the centre of retrofitting design, may need to change their working practices to ensure that sustainability issues are integrated into the whole design process. Sustainable input must not be left to one person or to occasional Building Research Establishment Environmental Assessment Scheme (BREEAM) or Leadership in Energy and Environmental Design (LEED) assessments; whole team reviews are recommended. Chapter 6 describes how one large practice has done this and put into place a system comprising a Green group and a 'toolkit' containing guidance, design prompt spreadsheet, assessment guidance and a graphic output. The toolkit covers all the sustainability issues and uses a 'rose' diagram to assess the achievement on each issue as the project progresses, which can be used within the design team at review meetings and with outside consultants. The rose simplifies sustainability performance to a graphic traffic light system. An additional use of the rose is to demonstrate to clients and agents the positive achievements in sustainability terms of the developing project.

Comfort is one of the important sustainability issues assessed by the rose and an important issue to all concerned with sustainable building retrofitting. Chapter 7 looks at how comfort can be improved during retrofitting without compromising energy performance. Analysis of comfort in buildings traditionally leads to the idea that thermal neutrality should be the target, giving rise to fixed temperatures to be achieved. However subsequent research led to the realisation that people happily adapt to a wider range of temperatures and this gave rise to the development of adaptive theories of comfort which when applied to offices particularly, could lead to much greater flexibility with consequent reduction of energy by reducing winter set temperatures and increasing summer set temperatures. Adaptive theory also applies to lighting where occupants can adjust to different light levels given the right conditions allowing greater use of daylighting, with the consequent advantages. Adaptive opportunities can relate either to individuals, such as office dress code, or to the building, such as openable windows and controllable blinds, and research suggests that making such opportunities available has a positive effect on overall occupant satisfaction.

Many refurbishment actions can directly affect comfort: insulation can reduce local under-heating and cold radiation, window and shading design affects glare, materials choice can cause off-gassing, exposing thermal mass can affect acoustics, and naturally building services and controls have significant effects on comfort. The designer needs to take an integrated approach and realise that most actions can have both positive and negative effects on the indoor environment.

Nevertheless, minimisation of energy use and CO_2 emissions in a refurbishment project must be a major part of the sustainability strategy. Much can be done to all existing buildings to make them more efficient, and decisions taken on most aspects of retrofitting design will affect the embodied energy and/or the final energy consumption of the building. Chapter 8 summarises the most common energy related refurbishment techniques and technologies and demonstrates these by links to case studies. Retaining as much of the building as possible rather than demolishing or removing it, and using natural and recycled materials, will reduce the use of energy in the construction. Facade design and construction, as in Chapter 9, will have a large effect on energy use in most buildings, as can insulation of structural elements, floors and roofs. Choice of ventilation strategy and related cooling methods are central to energy use and the management thereof. Roofing in courtyards to make atria provide both usable internal space and reduced heat loss has great effects on ventilation. Exposing or adding usable thermal mass is commonly cited as part of a cooling strategy to reduce active cooling, and heating systems and fuels are still important. Controls, both manual and automated, play a major role of all energy management. Connecting into local heat providing networks and collecting and using renewable energy sources should not be forgotten.

Facade design has always been a major factor in buildings for both aesthetic and practical reasons. The facade of a building is the essential element that largely determines internal performance but also the image of the building to the outside world. When refurbishing buildings, decisions on work on the facade are of primary importance. Chapter 9 looks at the options and issues involved for treating existing facades, related to their condition, image, performance and planning restrictions. Facade treatment can vary from major replacement, through refurbishment to only minor repairs. There are however many common issues that have to be considered, including considering natural ventilation, controlling solar gain, winter ventilation, insulation and cold-bridging, and surface temperatures.

Assessing and publicising the sustainability achievements of a retrofitted building is very important to all participants. 'Greenwash' is an inevitable consequence of the increasing public awareness of the importance of sustainability as the industry scrambles to demonstrate its positive involvement and progress being made. Environmental assessment is a useful tool for setting environmental targets for planners and clients, as check lists for design teams and on site, and for publicity and promotion of the finished building. From the early days of the development of BREEAM, the serious environmental assessment schemes have tried to measure real achievements, to raise standards and give credit where real sustainability has been enhanced across a wide range of issues. Chapter 10 discusses the main assessment methods that are applicable to refurbishment of non-domestic buildings. Some of these are very country-based whilst others are used almost worldwide.

Modelling of energy use in the completed building is an essential part of the design process for a sustainable refurbishment. Chapter 11 provides an overview of the modelling process and the range of tools available to assess both energy and related aspects of comfort. With an existing building the modelling can start with the plans and the components that could remain in the refurbished building and an energy model built to show the basic energy performance against which changes and improvements can be demonstrated. Energy use data for the original building may be available and are very useful if much of the building is to be retained and the post retrofitting use is likely to be similar, for calibrating and validating the model. Many different energy modelling tools are available and in use and they range from the more simple steady state, to complex dynamic models. Similarly comfort modelling tools range from simple models based on Fanger's research, to computational fluid dynamic (CFD) models that provide a view of conditions across a three dimensional space.

The energy model is used to compare different design options and packages of energy measures and to optimise theoretical energy use as the design progresses. However the model only ever estimates energy use and there is inevitably a 'performance gap' between this and the actual use when the building is occupied, due to a range of factors. Methods are being developed to reduce the performance gap during modelling, to produce more realistic consumption data, obviously important in influencing design choices for the better.

Energy may be the most important of the environmental sustainability issues but much else can be addressed to make the building more friendly towards the environment. The environmental assessment tools measure an increasing range of these issues and Chapter 12 describes these and discusses how they can be incorporated into sustainable retrofitting projects to best effect. Water use in buildings is responsible for source depletion, energy use and treatment issues and subsequently for waste disposal issues. The quantity of water used in buildings can be greatly reduced by retrofitting of water using devices, by better control and by collecting and using rainwater and even 'grey' water. Designing in proper waste storage and recycling facilities in and around buildings is also important. Guarding and improving landscaping and ecological facilities on and around buildings being refurbished can improve local comfort and counter the local heat island effect and can usefully be combined with sustainable drainage systems for local watering and reducing run off, easing pressure on drainage networks and reducing flooding risk. Materials for the refurbishment should be chosen on environmental grounds and much guidance is now available to achieve this.

Clients, governments, and society are looking to the construction industry to meet increasingly challenging targets of sustainability, energy performance and user comfort. Unfortunately, the construction industry frequently does not have the right experience and structures in place to deliver these improvements reliably. Following on from design, on-site construction needs to have sustainability embedded in the process as described in Chapter 13. This starts with the tendering process and choosing a contractor who understands the challenges of sustainable refurbishment. The contract may require more flexibility than a new build contract to accommodate unforeseeable situations and should include relevant sustainability performance indicators. Aspects such as demolition, asbestos removal and party wall agreements are important and controlling noise and waste disposal may be more difficult than with a new build project. A system for monitoring sustainable practices on site is needed and combining commissioning in with the construction phase will enable a smoother handover on completion.

However this leaves the underlying problem that designers and builders have been traditionally employed to produce or to alter buildings and expected to go away as soon as the work is physically complete and handed over. They are seldom asked or paid to follow through afterwards, to pass on their knowledge to occupiers and management, or to learn from the interaction. Consequently, the industry does not unlock all the value in the buildings it creates. The rigid separation between construction and operation means that many buildings are handed over in a state of poor operational readiness and suffer a 'hard landing', particularly – as often happens – when delays have led to the telescoping of the commissioning period.

It is also clear from monitoring of energy use in buildings that the occupants and facilities managers have a major effect on consumption. The reasons for this include poor building design, lack of understanding and mis-setting of the controls, insufficient facilities management input, and lack of control over out of hours use, etc. Designers can have some input by enabling easy and good control, provision of usable information and end user education and training.

Chapter 14 describes the comprehensive 'Soft Landings' approach to achieving occupational performance in line with design intentions. Although this chapter comes after the sections on design, modelling, assessment and construction, the principles and proposals suggested need to be seen as background to most of the intermediate chapters. We are not only talking about an integrated design process but an integrated design and use process in which designers think as users, and users and managers understand designer's intentions and act to achieve common sustainability goals.

Figure 0.1 summarises what can happen to energy performance from the strategy, to in-use stages in the building's life. Even at the strategy stage, where low energy principles are adopted, the unregulated loads can dominate and start the trend to high energy use. Project teams should work to solve the problems for and with the client, not pass the responsibility to the constructors or the building operators. The client also has to take responsibility for changes of mind, for not communicating changes and for not employing people with the right technical and management skills to operate the facility. The performance gap is therefore rarely a pure design gap.

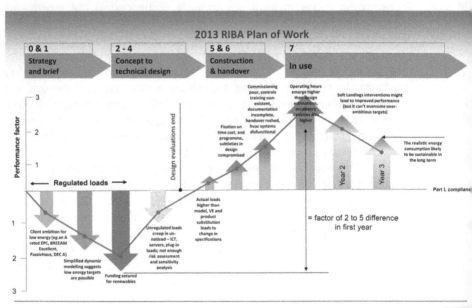

Figure 0.1 The performance gap

Source: BSRIA

THE CASE STUDIES

The chapters described above talk about the why and how of sustainable retrofitting and what can be done to support this within the whole industry. The case studies in Chapter 15 have been chosen to show what is actually being done and what is being achieved in a range of buildings in different countries. The approach taken, the technologies used, the degree of rebuilding, the success and conclusions all vary, as they inevitably will in different situations. The message is that old buildings are being reused, adapted to new uses, retrofitted with the latest technologies, and brought up to modern standards and to current expectations. This in itself is likely to be more sustainable than demolition and new build, but there is much more that can be done to create truly sustainable commercial buildings, highly energy efficient, comfortable for occupants, using 'green' materials, minimising water and drainage use, minimising waste, including renewable energy sources, etc.

1
DESIGNING FOR NEW USES, STANDARDS AND REQUIREMENTS IN THE TWENTY-FIRST CENTURY

John Davies

For developers as space providers, a key driver is to provide well-designed spaces that people want to occupy. In doing this and to serve the market well, it is important to understand what a well-designed space consists of and what the customer wants and expects. Only by knowing these things can we successfully address the issues which surround sustainability in the refurbishment of commercial property, as ultimately they represent the key areas to address in ensuring a truly sustainable approach.

When looking at the design and refurbishment of existing buildings and spaces, a number of challenges are presented, not least that there is no 'clean slate' to start from. The lack of a truly blank canvas means economies of scale are sometimes difficult to achieve, e.g. the purchase and availability of certain materials and systems or the simple availability of development space to work with. Likewise achieving a higher level of energy efficiency from ageing plant and systems is very often impossible without some kind of significant replacement. As a result refurbishment is quite often seen as a hassle when compared to starting anew. This in turn leads us to a wider issue that we in the UK are beginning to witness, in that there is an apparent shortage of design expertise when it comes to the refurbishment of existing buildings, particularly refurbishment in the aged property sector. Whether this shortage is a reflection of the many years of growth in the UK property sector and the prominence of new build as the default option, or that refurbishment is simply not seen as a speciality for designers, it is clear we are simply not seeing refurbishment as a first choice option when compared to new build.

However, taking these challenges into consideration (perceived or otherwise), do they really inhibit us in taking on refurbishments on a more consistent basis? Indeed, are we missing an opportunity to create more sustainable buildings by retaining and regenerating older ones? Moreover, are we likely to reach a point whereby the preferences for new build versus refurbishment will be reversed as the existing building stock continues to grow and a new status quo is established? Whilst we cannot categorically answer all of these questions, we can see changes and shifts in preference beginning to occur. This is very much the case in the context of London, whereby the vast majority of the building stock is aged and/or existing, with unique historic profiles to many of its villages and buildings. Coupled with this, using London as an example, our experience tells us that in many cases it can be more advantageous to refurbish as it can often be quicker, both in terms of programme and planning, and cheaper than redeveloping from scratch. Moreover, from a resource efficiency perspective it can be more sustainable, as you are potentially retaining and reusing more of the building.

Location has always been one of the critical 'umbrella factors' in property. It often motivates a customer's decision in taking space – for which there are many drivers, e.g. wanting to be near a customer base or near to the competition, and it often sets the rental income guidelines for a given area. However, there is also an element of tradition when exploring the location factor. Many customers will remain in an area for decades as it is seen as the right or traditional district or area for the location of their given business operation. This in turn gives rise to areas dominated by particular business trades, e.g. lawyers and financers, and as a result may have a particular architectural theme or bias which gives an area its character or identity; a trait reflected in many cities around the world. However, removing locational desire expressed by customers from the equation, a building's actual location can play a key role in how one approaches its refurbishment; for example, considering transport and connectivity, how to incorporate the additional levels of infrastructure that may be required to create the right level of integration into the surrounding public transport networks. Can the building sustain these? Likewise, can space allowance be made inside the building for other modes of transport, for example bicycles, and the associated facilities for their riders? All these factors can drive the decision as to whether a space provider refurbishes or redevelops.

Coupled to all of this is the end user or customer. Understanding their needs from the outset is critical in achieving the right outcome and space solution. However, you are not always guaranteed to know your customer, their business or their requirements, so being flexible and agile to your market and responding to occupation patterns and preferences is paramount. In recent years we have seen a clear shift here in the UK of customers wanting to temper space costs in a hardening market, whilst making the most of technological advances to improve business efficiency. This has seen many businesses reposition assets and space in their corporate portfolios by co-locating business functions and teams and using denser occupation strategies – 'max packing', typified by agile working, hot-desking, touch-down areas and mobile workstations. However, in the process of condensing teams and workspaces, there has also been interest raised as to whether a space is able to promote and enhance a higher level of employee well-being; and, should denser occupation strategies be employed, whether the space itself can help with making the overall environment a healthier or better one to work in. Further to this we are also seeing changes in how businesses are 'using' buildings and seeing them not just as places of work but also as an asset by which to attract and retain employees. Likewise, businesses are choosing buildings in less traditional locations with less conventional features in order to develop a unique selling point to their business.

The following sections explore the above issues and seek to recognise how the role of good design in refurbishment coupled with an understanding of changing customer requirements, whilst using the under-lying principles of sustainability, can help create the new, exciting but above all efficient spaces of the future.

THE VALUE OF SUSTAINABILITY

Sustainability is an important and valuable tool for property developers and investors as it provides a framework which helps define approaches to efficiency and better practice, but also helps meet the needs of customers in a more proactive way, although this has not always been the case. For many years sustainability has been viewed as a 'nice-to-have' item, but not critical to the day-to-day machinations of the property industry. However, like so many things, this has begun to change. Sustainability is now widely seen as a 'must-do' or indeed a 'want-to-do', not just from a legislative perspective but also from a market/customer demand perspective – having a Building Research Establishment Environmental Assessment Method (BREEAM) 'Excellent' or Leadership in Energy and Environmental Design (LEED) Gold certified office is merely the starting point of the discussion these days. For those who have embraced it culturally and have worked out how to 'use' it effectively it is already reaping them the rewards – improved shareholder and analyst confidence, quicker lettings, stronger covenants and better long-term values, all of which are signs of a well-executed business model and approach to sustainability which enables organisations to demonstrate value. Whilst empirical research data is still sparse, those who have embraced it are evidently realising these benefits and believe there is a clear link between sustainability and value.

However, whilst we can clearly see and appreciate these benefits, how are they derived? So far as we can tell these value indicators stem from a mixture of customer and market confidence, but more specifically confidence stemming from the fact that the buildings and spaces (the product) on offer are gauged correctly for the market and fit with customer requirements. Furthermore, we are able to attribute much to the design of the building or space, as it is the design which encapsulates the various landlord and potential user requirements, conflicting or otherwise, and has to balance them in order to create the basis of the product. It is at this point that some see sustainability as a conflicting priority, often sidelined by cost-engineering exercises and caught in the inertia of whether it is worth investing in more efficient and sometimes more costly design approaches, in order to save money in operation and provide an enhanced product, than it is to simply get the product to market as quickly and cheaply as possible. However, it is important to take a macro view and one which focuses on the value of the asset married to an understanding of the market and what the customer wants. As a result good design should help to deliver a sustainable

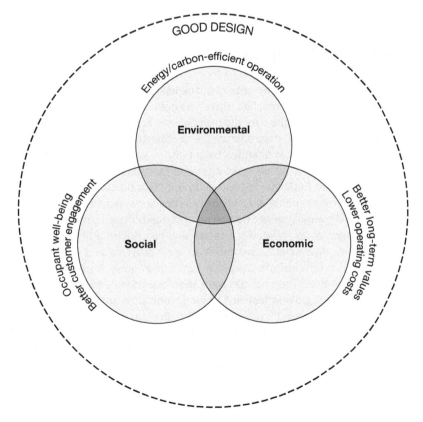

Figure 1.1 Relationship between sustainability in property and good design

Source: Derwent London plc

approach, moreover create a building or space which is ultimately desirable – remembering desirability sells! Furthermore, there should be no difference whether it is in a new build or refurbishment scenario – after all, good design is good design regardless of where or how it is employed.

So how does all this fit together? Where are the linkages? By using a classic Venn diagram we are able to show some of the key property-based aspects which are important to both developers and customers and where they fit in terms of the sustainability agenda.

Whilst Figure 1.1 is not exhaustive in terms of being able to describe every single aspect, it does highlight some of the key issues the industry is faced with which need to be addressed, and that good design in one form or another can help realise. However, these sustainability issues are by their very nature dynamic; they do not stand still and are forever changing, and their effects felt in different ways. Therefore, it is imperative to have in place a robust management approach to sustainability, which cascades into all business activity and ensures it is included in all decision-making. Only then can learning be captured which can then be used to adapt to future change, be that legislative, customer reaction or simple industry progression.

CHANGES IN THE LEGISLATIVE LANDSCAPE

Coupled with the rise in prominence of sustainability and organisations developing their respective approaches, there has been a considerable change in the UK legislative landscape over the past decade, in particular the legislation governing the environmental performance of commercial property.

The most notable changes have been in the Building Regulations and in particular Part L, which looks after energy conservation in buildings. This section of the Regulations has become a central pillar of the modern movement of environmental sustainability in buildings, exemplifying its rise with a rapid and ever-increasing tightening of the carbon performance criteria which buildings are to meet. As a result this has driven a change in the approach many developers have taken to their projects and in particular raised the bar for refurbishment projects which previously were not subject to such tight controls. The changes have seen buildings become 'tighter' in terms of their envelope permeability, heating, ventilation and air-conditioning (HVAC) systems becoming more efficient and diverse, e.g. mixed mode and controlled more sympathetically, and on-site energy generation from renewables more widespread. Whilst there have been cost implications as a result of these changes, many would argue that the Regulations have helped to deliver a higher standard and better quality of space which might not have existed without such rules. However, even then these Regulations are not static, far from it; they will escalate and get tighter. As a result only now having just adjusted to the latest version introduced in 2010 the next iteration for 2013 has just been announced, calling for further reductions in emissions.

As with previous incarnations we will have to wait and see how this will affect a given development and what the cost implications might be; however, it does provide much needed guidance on what is expected so that the property sector is able to contribute to the UK's legally binding 34 per cent and 80 per cent carbon reduction targets by 2020 and 2050 respectively.

In addition to building standards, the Energy Act 2011 is making its presence felt, bringing with it a series of historic landmarks which are set to provoke a change in the market. The first and most widely known change is the introduction of minimum energy performance standards for commercial property and the legal requirement for all lettable property to meet a certain standard from 2018 onward. Whilst the exact means by which this will be tested and what the actual minimum level will be are still not clear, there is firm evidence to suggest that the Energy Performance Certificate (EPC) will be the measurement tool, with 'E' rating being the minimum threshold. As a result landlords, developers and investors are rapidly assessing their portfolios to establish their compliance base and their subsequent risk mitigation strategies to ensure their spaces remain lettable beyond 2018.

Figure 1.2 provides a very brief snapshot of some of the pertinent regulatory changes we will witness across the property industry in the coming years, all of which culminating with the government's desire to have all new commercial property zero carbon by 2019.

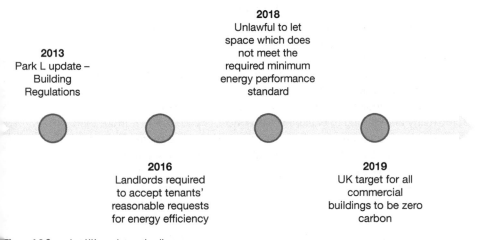

2013
Park L update –
Building
Regulations

2018
Unlawful to let
space which does
not meet the
required minimum
energy performance
standard

2016
Landlords required
to accept tenants'
reasonable requests
for energy efficiency

2019
UK target for all
commercial
buildings to be zero
carbon

Figure 1.2 Snapshot UK regulatory timeline

Source: Derwent London plc

UNDERSTANDING THE CHANGES IN BUILDING OCCUPATION AND USE

It is not hard to see that the modern workplace has changed compared to that of its predecessor of decades past. The macro changes in how companies approach managing and motivating their workforces, distinct shifts in working pattern and process, coupled with a desire to get closer to end markets and customers, has seen the modern workplace change rapidly. This has manifested itself in a variety of ways, and this has not happened in isolation; rather, it has been supported in part at least by significant advances in supportive technologies, allowing users to communicate more quickly and become more agile in the delivery of their work or services.

Interestingly this evolution is only now really starting to intrinsically affect the spaces that companies occupy and the types of spaces they seek, out of preference and/or business need. Previously we witnessed that some businesses simply try and adapt their immediate surroundings to fit this new world order, with varying degrees of success. However, companies are now more readily re-evaluating their space needs and asking themselves what exactly is important to them in terms of space provision for their business. Is owning a large headquarters still required? Is greater flexibility needed? Consequently, clear trends are emerging with companies wanting to do more to deliver better value for money from, and maximise the potential of, their space. Trends such as using denser occupation ratios or 'max-packing' when arranging their space layouts, are often seen when co-locating workforces from different buildings or floors, or locating to smaller spaces. Also buildings and spaces are being positioned not just as a place of work but as an integral asset designed to attract and retain a talented workforce. This new emerging focus is strong in sectors such as technology, media and telecoms (TMT), where traditional approaches to space selection and management, e.g. location, are being ignored in favour of spaces and layouts which are flexible, looser fitting and can be adapted to cater for the needs of the individual and ensure their well-being, such that they can deliver their work more effectively. A knock-on effect of this is the shift in working patterns. We are seeing employees spend longer in the workplace, as companies encourage greater amounts of collaboration and creative working, breaking traditional working patterns and extending the working day. Although this is not true of every sector, many are starting to explore these concepts and look to nurture their employees such that they can deliver at their best and be more productive. However, having said all this we are also seeing things which are not changing. Customers are still showing desire for the basics in terms of space prerequisites, i.e. lots of natural light, tall floor to ceiling heights and generous space volumes. Either way it is evident that there is a keen desire to do more with less, and continue to develop an ever-agile workforce, less dependent on fixed positions and more dependent on mobile operability.

Whether these trends are driven by simple evolutionary change or market forces (or both), there is a clear subtext, namely reduced operating cost per person, increased efficiency, better productivity and better well-being. This is not at all surprising as is it vital for a business to maintain a healthy balance sheet such that it can sustain itself for the future, but it is also vital to attract and maintain a high performing workforce. However, what is surprising is the impact this is having on the commercial property sector and the speed at which it is able to adapt and ultimately serve its market with what it needs, recognising that buildings are complicated and cannot simply change overnight to be something different. And it is these points that raise some important questions: are we responding to market need appropriately in terms of the spaces we are currently providing? Are they future-proofed and are they flexible enough?

Consequently, these important questions ask us to look at how we currently specify space, new build and refurbishment alike, whether this is sufficient to respond effectively to future need. In the UK, the principle approach to space specification follows patterns established by the British Council for Offices (BCO) guide to specification. These guidelines provide a range of good practice parameters, for example internal comfort conditions of 24°C for cooling and 20°C for heating and occupational densities of one person per 10 m^2. Whilst these parameters serve as a valuable aide-memoire and sense check when designing space, they have risen in prominence to become a de facto standard. Such is their prominence that they have become ingrained in many property professionals' approaches and management techniques. For example, property agents use the parameters as a key tool during their sales and letting processes as an assurance factor for customers who are seeking confidence that a space conforms to a recognised standard. However, is this confidence and uniformity allowing us to move effectively with the evolution and changes in occupation and use patterns we are seeing? Some might argue that this uniformity is creating rigidity which is not allowing newer approaches to come forward.

Another aspect to explore whilst examining these changing occupation trends is the knock-on effect in terms of how efficiently a space can perform when it is more densely occupied. More people in a space will mean more cooling is required during hotter periods; likewise there is likely to be a greater draw on plug loads to cater for increased small power requirement and more lighting. This in turn can lead to higher operational energy profiles and carbon footprints, especially if a space has been designed to a person/m^2 ratio of 1:10 and it is being occupied in reality at a ratio of nearer 1:8. This is further compounded when customers may want to augment or change a space, e.g. add partitions and break-out spaces, which may not have been included for in the base design assumptions and as such compromise the servicing strategy or structural grid. Therefore, to try and counteract or balance this more effectively, it is important to understand the realities of how space is actually occupied.

So how can we marry the apparently differing perspectives of the current approach to space provision and changing customer requirement? As mentioned earlier, good design plays a pivotal role, and ensuring the right 'people-centric' design aspects are included in a base design is key in ensuring a space is as effective as possible. Whilst it could be argued that this is a fairly straightforward task when presented with a blank canvas such as a new build, is it that simple when undertaking a refurbishment? Interestingly, in some circumstances it is, particularly when dealing with major refurbishments which involve stripping a building back to its bare frame. This can allow the introduction of new floor configurations, different, higher quality, light penetrations and newer more efficient building service technologies. However, historic buildings such as old industrial buildings often present themselves initially as unlikely candidates in terms of capturing many of the well-being factors sought, such as large and open space volumes, generous floor to ceiling heights and lots of natural light. But many do enjoy a lot of these features as well as high levels of thermal mass which, if harnessed correctly, can create a very energy efficient, passive strategy to heating and cooling. However, that is not to say that all these are present in one building all of the time, but given the breadth of the existing commercial building stock, particularly in London, there is a significant proportion of buildings which could fulfil these criteria.

As well as potentially providing some of the key well-being characteristics, some older and historic buildings also present good scope in terms of flexibility and change of use capability, lending themselves to a wide range of applications, aside from their original intended use. This versatility allows us to exploit a building again, in whole or in part, thereby avoiding the environmental, capital and time costs associated with its demolition and those of starting the build process from scratch. Embodied carbon studies show that building refurbishments can reduce carbon footprints by as much as 70 per cent when compared to new build solutions. Obviously much of this benefit is derived from retaining key elements such as the foundations and superstructure, but it shows at a practical level that major carbon savings can be achieved by looking at buildings in a slightly different way and using good design and market intelligence to find potential and practical solutions in sometimes the most unlikely of places. In addition to the obvious environmental gains, retaining existing buildings also assists with place-making and heritage retention. This can be very important in the context of historic districts and conservation areas within towns and cities and ensures that buildings remain in situ, albeit in perhaps a slightly revised function or, indeed, form.

ADAPTING TO CHANGE: TEA BUILDING, LONDON

An example of this change of use, and how a historic building can be brought firmly into the modern commercial building stock, the Tea Building in Shoreditch, London demonstrates transformation from factory/warehouse to cutting-edge office space.

The Tea Building (Figure 1.3), an imposing eight-storey building, is a landmark at the junction of Shoreditch High Street and Bethnal Green Road, London. Built in the early 1930s for Allied Foods' Lipton brand as a bacon curing and packing factory, it is joined internally to the Biscuit Building next door, a slightly older but equally impressive warehouse which was principally used as a tea packing warehouse.

gure 1.3 The Tea Building
ource: Derwent London plc

As part of its transformation, the 26,000 m^2 building underwent extensive improvements to its layout and servicing in a series of development phases to create a number of dynamic and modern spaces which include cafes, commercial offices, galleries, clubs and a hotel.

Focusing on the commercial office space, the spaces are self-contained studios, all of which are different shapes and sizes, housed within the main floor plate, and planned around broad internal 'streets'. In 2010, as part of the ongoing refurbishment programme, the 'Green Tea' concept was introduced. This set in place a protocol of upgrades to the spaces as they became vacant, improving the glazing, lighting and heating and cooling, such that their operational performance is significantly improved, as well as introducing a building-wide thermal loop on the roof designed to provide heating and cooling based on the position of the sun.

RESPONDING TO CHANGE – FUTURE SPACES

As the market continues to adapt to ever-changing economic conditions, it is vital that space providers respond to the change in customer requirements and the shifts in terms of space selection and occupation trends, so that spaces are future-proofed. A key aspect of this is flexibility, and this comes in many forms. Spaces should be attractive to a wide range of businesses and customers who should be able to easily make it their own, but also the space should be adaptive, i.e. having the ability to switch between uses easily without major impact on the overall management strategy of the building.

Coupled to flexibility, future space provision will need to be technologically 'intelligent', i.e. allowing businesses to maintain their virtual location and presence via super-fast internet connectivity, vital for the dot.com sector. Also spaces must have 'smart' controls and business support systems integrated into their fabric to allow efficient control and management, e.g. energy monitoring and comfort control. However, this should not mean over-specification. Rather, by taking a leaner, simpler and looser fit approach it is more than possible to incorporate the much-needed flexibility and technological features. Indeed in many cases it is easier, especially in refurbishment scenarios, as it can be difficult to incorporate the traditional 'necessary features' such as suspended ceilings into older buildings successfully without compromising overall space provision. Therefore, the confines of a refurbishment project can engender a somewhat fresher approach and one which is more considered as opposed to generic. Moreover, there are also sustainability benefits, outside the obvious retention of building features. A looser fit and simpler approach to design specification can lead to resource efficiency savings by simply not specifying the additional levels of fit-out materials and equipment required in the base build, thereby lowering embodied carbon profiles and reducing waste generation.

As well as a simpler approach to design, a simpler approach to operability will also be necessary, together with a greater level of connectivity for occupants with the building. Whilst we are seeing the introduction of ever more complex and sophisticated building management systems (BMSs) which are capable of measuring and manipulating remotely and automatically many features within a building, we are also seeing a reduction in the level of occupant interaction and connection with a building. Although BMSs are important and they create efficient operating environments, they may not be completely in tune with occupants. For example, a BMS does not react to individual preferences on temperature satisfaction and know from an individual's perspective when a space is too hot or too cold or when an individual would like the lights on or off, or a window open. It can only work by the predefined set points and ranges set within its programme routines across a selection of predefined zones within a building. However, it is often cited that the reason for this increased level of automation and control is to reduce the 'user-error' factor and to prevent misuse of a space by its occupants, which leads to inefficiency and waste. Indeed, it is true that occupant behaviour within a space is a leading factor in terms of its efficiency, and how much energy is consumed. But is more complex automated control the right way forward? Moreover, what is required in the future? Perhaps what is needed is a more integrated approach, one which shares both the benefits of a BMS coupled with occupant-initiated demands, i.e. a BMS that adapts to the choices made by the occupants, and likewise suggests the best course of action for a given situation in conjunction with the user. For example, occupants could tell the BMS, via a 'smart device' such as a phone or tablet, that they are too hot or cold in a given area, and the BMS could provide an option or series of options to adjust the situation, as opposed to just simply changes based on preset points.

THE FUTURE AND LEARNING FROM OUR PAST: WHITE COLLAR FACTORY, LONDON

Drawing all these aspects together it is clear that we need to adapt and make our spaces more flexible, and learn the lessons not only of the past but also the present, distilling the best approaches and techniques into designs in order to craft our future spaces; a generic tick box approach is simply not good enough. The White Collar Factory development (Figure 1.4) encapsulates both widespread refurbishment and new build elements, and incorporates occupant-centric features and sustainability attributes which show how the old and new can be brought together to create flexible, future-proof spaces.

Derwent London undertook several years of research examining exactly what prospective customers want when selecting space, what motivates them, what features are seen as 'must-have' and what are desirable. The White Collar Factory campus development embodies the culmination of this research and sets out an innovative, highly flexible and efficient space arrangement. It is located in 'London's Tech Belt' on a corner of the Old Street roundabout and is comprised of six buildings, five of which are existing buildings which will undergo extensive refurbishment and the remaining building being new build which will be a twenty-first-century interpretation of the industrial buildings of the past. Whilst this particular building in the campus is not a refurbishment it incorporates many features that are present in older, more traditional buildings, and makes a clear departure from the current modern commercial stock.

Figure 1.4 White Collar Factory, London
Source: Derwent London plc

END PIECE

Trying to map and understand the future market trends and space needs is not easy and there are no magic formulas or equations; rather, it relies on developers and landlords spotting trends and understanding their customer bases. However, it is safe to say that our space provisions will change and continue to do so as the modern workplace evolves and businesses introduce ever more diverse ways of managing their workforces and product delivery.

To respond and deliver to this change, good design needs to be effectively deployed to capture the needs of the customer within a space, and ensure that it is as efficient as possible to run and flexible to occupy. Moreover, this can be achieved relatively easily in a refurbishment scenario and, in many cases, create a product which is as equally desirable if not more so than a new build solution. However, it is important we learn from the buildings of the past such that we can capture their best attributes and distil these into our future spaces – one could say we need to go Back to the Future!

2

VALUE AS A DRIVER FOR SUSTAINABLE REFURBISHMENT OF COMMERCIAL BUILDINGS: A EUROPEAN PERSPECTIVE

Ursula Hartenberger

BACKGROUND

Recent reports, studies and literature reviews point to a growing body of empirical evidence that high-performing commercial buildings not only perform better in terms of resource efficiency but also financially in terms of commanding higher rents or resale prices, higher occupancy levels and increased employee productivity and well-being. Research has also shown that not only residential users might be more willing to pay for sustainable homes (Van Eck, 2008) if they are offered to them in a way that offers value for money, but that there is now a growing group emerging from within the investor and corporate occupier stakeholder communities who are also increasingly opting for more sustainable buildings on grounds of better financial returns in whatever shape and form they might be strategically important for them individually.

However, as promising as all this may sound, workshops with property owners[1] in Europe have highlighted that there still is a prevailing lack of confidence regarding a reasonable return on capital in the market which is also highlighted in a report by the Buildings Performance Institute Europe (BPIE, 2013) according to which the monetisation of the benefits that arise along the energy cost savings is frequently overlooked. Therefore, a vital aspect in making the business case for investing in sustainable refurbishments of any building – whether residential or commercial – is the building's potentially increased future value. This value is usually determined through a valuation of the property which is either established at the point of transaction, e.g. when buildings are being sold or rented or in the case of property held in investment funds at quarterly (or even monthly) intervals regardless of their status and whether they are for sale or to rent. Therefore, in order to raise awareness on the value of investing in sustainable refurbishments of existing commercial buildings amongst prospective buyers, sellers, lenders and investors, incorporating sustainability considerations into valuation practices is absolutely essential.

Against this background, this chapter will give a brief overview of the existing literature surrounding the business case for sustainable commercial real estate, examine the role of valuation in the wider economy and in transforming commercial property markets as well as current market barriers to stepping up the pace of investing in sustainable refurbishment of the existing commercial real estate stock with regard to valuation and how these could be overcome. Finally, it will also look at different types, perceptions and concepts of value that can act as a driver for either investors to underwrite investments for sustainable refurbishments of commercial buildings or for certain groups of occupiers to want to occupy these buildings.

EXISTING LITERATURE SURROUNDING THE BUSINESS CASE FOR SUSTAINABLE REFURBISHMENT INVESTMENTS

Since 2009, a number of economic impact studies have been undertaken by a variety of different stakeholders, most recently by The Institute for Market Transformation in the USA (Sahadi et al., 2013), the World Green Building Council (WGBC, 2013) and the European Commission (Bio Intelligence Service et al., 2013), the latter looking at Energy Performance Certificates and their impact on transaction prices in a number of European Union member states, regions and cities. Apart from this Commission report, the majority of studies conducted to date mainly refer to US data sets with the exception of two studies commissioned by the Royal Institution of Chartered Surveyors (RICS) investigating value premiums for sustainable buildings within the London office market (Chegut et al., 2012) and the potential economic impact of Energy Performance Certificates within the residential sector in the Netherlands (Brounen and Kok, 2010) respectively, in addition to a few national or regional studies conducted in Switzerland (Salvi and Syz, 2010) and Germany (Wameling, 2010) and finally a study by Kok and Jennen on the value of energy labels in the European office market (Kok and Jennen, 2012). Unlike in the USA, widespread systematic data evidence on financial performance and value enhancement through sustainable refurbishment investments is still not readily available in Europe (see section on Current Market Barriers below) and whilst the US-based studies certainly help to make the overall business case for sustainable buildings, they are not of much use to the local valuer who would need local data to carry out a meaningful valuation.

THE PIVOTAL ROLE OF VALUATION

The use of property valuation to provide a (market) value is an essential aspect of the property life cycle. Accurate valuations are important for a healthy property market and ensure stability in both developed and emerging economies, forming the basis of performance analysis, financing decisions, transactions, development advice, dispute resolution and taxation along the whole property life cycle as illustrated in Figure 2.1.

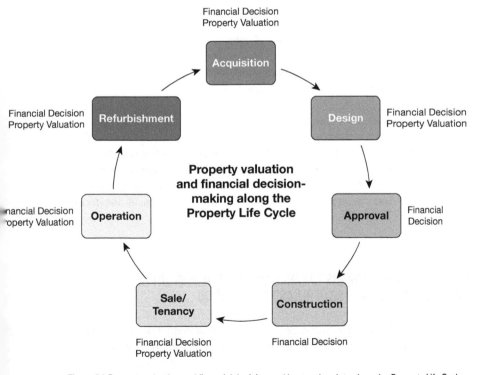

Figure 2.1 Property valuation and financial decision-making touch points along the Property Life Cycle

Source: RICS

There are several definitions of value underlying commercial property valuations. The two most common definitions (from the International Valuation Standards) are:

- Market value – the estimated amount for which an asset or liability should exchange on the valuation date between a willing buyer and a willing seller in an arm's-length transaction, after proper marketing and where the parties had each acted knowledgeably, prudently and without compulsion.
- Investment value – the value of an asset to the owner or a prospective owner for individual investment or operational objectives. Differences between the *investment value* of an asset and its *market value* provide the motivation for buyers or sellers to enter the marketplace. This is also known as 'worth' in the marketplace.

Both definitions provide the basis for an estimate of value at a certain pinpoint in time and this is particularly valid at the point of transaction, e.g. when buildings are being sold or rented, as this is usually a period when specifications for future fit-out and refurbishment investments would be considered. The extent and approach of reflecting sustainability in value estimates strongly depends on:

- the underlying definition of value;
- property type;
- regional and local market conditions;
- regional and local conventions, etc.

Looking at figures from the UK Office of National Statistics, there were 97,670 transactions in the commercial property market segment in the UK for 2012, underlining the potential of triggering renovation at transaction stage very poignantly as all these transactions will have required valuations, stressing not only the importance of intervention at transaction stage but also of collecting and managing the transaction data as a basis for future valuations.

CURRENT MARKET BARRIERS TO LARGE-SCALE REFURBISHMENT OF COMMERCIAL BUILDINGS WITH REGARD TO VALUATION

Valuers are well placed to comment on the financial impact of asset specific and wider market factors and as demonstrated above, the point of valuation and the subsequent financial decision-making present opportunities to consider more sustainable design options. Therefore, as current valuation techniques have the capacity to reflect sustainability features, harnessing valuers' expertise in the drive for improving the sustainability performance of commercial buildings is absolutely pivotal. However, in terms of seizing these opportunities there is a set of interlinked barriers.

Inadequate data and information management sources

Data availability and transparency and information flows are of key strategic importance when valuing a building, advising clients and ultimately trans-forming the market. Valuers use a number of data sources when calculating 'Market Value'. And yet, one of the main challenges for valuers, particularly in Europe, is access to good quality building performance and transaction data that would help to define and communicate true costs and savings potential of implementing sustainability measures and forecast potential future risks and opportunities for both commercial property portfolios and individual assets. Valuers face a difficult situation when assessing 'Market Value' as they must reflect current market demand and as such they cannot by default 'make' the market for sustainable refurbishment of commercial buildings. Without existing (local) market evidence they are not in a position to consider sustainable building features as part of their standard valuation practice. This creates a vicious circle whereby clients are not adequately advised about the value and long-term benefits of sustainable solutions and so investment and demand for these remains stagnant. The current lack of transparency within property markets, mainly caused by the rather unsys-tematic gathering and management of both actual building performance and market transaction data, remains the most significant barrier to understand-ing the link between its sustainability performance and a property's value.

In terms of market transparency, even in a country like Germany, a country often quoted as an EU best practice example when it comes to policies and financing tools, transaction data is not captured in a central database according to standardised criteria, but often stored in different locations using different computer systems and data formats. As a rule, the information from energy performance certificates is not systematically linked to transaction data either. Both transaction- and building-specific information is of varying quality due to a lack of communication between the different valuation committees holding the data who also may not always see the relevance of adequately managing sustainability related data. This in turn makes it difficult for valuers to factor sustainability into their valuation reports. This is more or less exemplary of the current situation in most other EU countries. Driven by the concern that, despite legislation, the market did not appear to place value on energy efficiency measures, the UK government commissioned a study in 2009 (RICS, 2009) to help understand the lack of demand for energy efficient (residential) property and to come up with practical recommendations for government and the property sector on how to overcome this lack of demand by increasing the importance of low-carbon buildings and technologies in relation to the market value of property. One of the key findings was that improved information and information flows were at the heart of a successful transformation to a low-carbon built environment. The importance of having reliable data in the decision-making process in favour of opting for sustainable refurbishments was also highlighted by the aforementioned 2013 literature review carried out by the World Green Building Council on the business case for green buildings which noted a marked 'perception gap' between the perceived (0.9–29 per cent) and the actual (-0.4–12.5 per cent) cost premiums for green buildings at design stage (World Green Building Council (WGBC), 2013). This somewhat distorted perception that 'going for green' automatically equals substantially higher costs can only be successfully addressed if the sector brings its data management house in order. Commercial property owners and investors need to be provided with more qualitative advice about 'cost-return-value-impact' scenarios to incentivise them to opt for performance improvement measures. New ways of gathering, processing and presenting property-related information by all stakeholders are required. This is particular the case for property transaction databases which should be extended to include building performance data. Valuers should collate sustainability-related information when carrying out valuations even if the data may as yet not be value-determining. And, finally, the onus should not only be on the valuer to source the client's data. Building owners and managers also need to apply due diligence with regard to their own data management and start perceiving data as an asset in itself as going forward; inadequate data management on the part of the client may well find its way into valuation reports as a risk factor.

Knowledge, skill and awareness gaps

Professional guidance on how sustainability can relate to a built asset's value started to emerge at the end of the last decade at a time when the so-called 'green building' movement was gaining momentum.[2] Whilst current valuation techniques have the capacity to reflect sustainability features, at present there are still quite significant knowledge and skill gaps amongst valuers when it comes to understanding specific user needs and the possible value impact of existing and emerging sustainability technologies. It would be unfair to blame valuers for not holding the capability to assess these technologies as this type of knowledge traditionally falls more in the domain of building specialists such as architects, building controllers, building surveyors and facility managers. Nevertheless, whilst there may be variances in some markets, it is safe to say that at this point awareness and understanding amongst most commercial valuers about the implications of the aforesaid studies as well as of their own role with regard to considering sustainability in valuations is still relatively low, and the lack of technical knowledge can make it very difficult for valuers to translate the often more technical and granular information into more aggregated board room financial decision-making language. Better collaboration and communication between professionals representing the more technical and financial aspects respectively would be needed to address this because if the majority of valuers were able to offer their clients evidence-based advice in addition to their customary reporting services during the transaction phase this could have a significant market impact.

Lack of engagement of lenders and providers of finance

Lenders are arguably one of the key market players within the valuation process. However, lenders' business models are obviously driven by commercial returns and prudent lending decisions. Whilst initiatives driven by the Sustainable Building Alliance,[3] the Property Working Group within UNEP FI[4] and UN PRI[5] signatories as well as a recent report by the Institutional Investors Group on Climate Change[6] (IIGCC, 2013) illustrate that leading European investors are now beginning to recognise the economic impacts from investing in green building and that this changing mindset is now also beginning to affect the property market, at present, most European lenders are not really actively engaged in the debate around sustainability and value in the built environment. Banks are typically not used to assessing the risks and benefits of sustainability-related investments and may therefore be reluctant to underwrite financing them. In the same way that consumers and private homeowners will favourably respond to sustainable solutions, lenders and investors might feel more inclined to do so if the business case was presented to them in a language that they can relate to. In most cases they do not understand the granular technical data relating to the potential performance of the latest state-of-the-art technology, and the more technically inclined stakeholders may well need to accept that they do not actually need to. Most of the environmental metrics currently on the market were developed without either considering the needs of the investment community nor were investors involved with UNEP at conception stage. This is underlined by a report by the Property Working Group within UNEP FI (United Nations Environment Programme Finance Initiative, 2011) which points out that issues such as environmental metrics tend to be perceived by investors to be more of a technical nature and, therefore, of limited direct relevance to their day-to-day decision-making. Given their role as information managers, valuers would be ideally placed to bridge the gap between technical and investment data analysis.

In conclusion, this section has shown that whilst awareness and engagement may be on the rise amongst key stakeholder groups, at the same time there is also no shortage of barriers to large-scale sustainable refurbishment of commercial buildings in Europe as summarised in Figure 2.2. Essentially, these barriers mean that the potential that both the legislative, such as the EU Energy Performance of Buildings Directive (European Commission, 2010) and existing voluntary frameworks offer has not been fully seized.

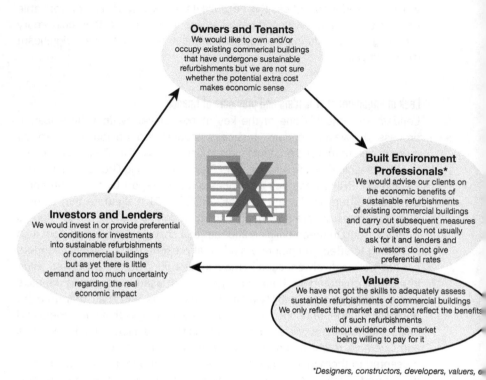

Owners and Tenants
We would like to own and/or occupy existing commerical buildings that have undergone sustainable refurbishments but we are not sure whether the potential extra cost makes economic sense

Built Environment Professionals*
We would advise our clients on the economic benefits of sustainable refurbishments of existing commercial buildings and carry out subsequent measures but our clients do not usually ask for it and lenders and investors do not give preferential rates

Investors and Lenders
We would invest in or provide preferential conditions for investments into sustainable refurbishments of commercial buildings but as yet there is little demand and too much uncertainty regarding the real economic impact

Valuers
We have not got the skills to adequately assess sustainble refurbishments of commercial buildings We only reflect the market and cannot reflect the benefits of such refurbishments without evidence of the market being willing to pay for it

*Designers, constructors, developers, valuers, e

Figure 2.2 Current market barriers to large-scale uptake of sustainable refurbishment of commercial buildings
Source: RICS

VALUE-DETERMINING DRIVERS

If value is a potential driver for sustainable refurbishment, the question arises as to what would be the specific sustainability relate value-determining drivers. A 2012 Austro-German-Swiss sustainability-focused guidance for valuers (Meins *et al.*, 2011) takes a novel approach to featuring key value-determining drivers in a so-called 'long list', fully integrating sustainability aspects into a comprehensive checklist consisting of twenty-three characteristics with around eighty individual indicators categorised into six overarching categories: location, site, building, economic, image and quality of processes. The objective of the Neuer Bewertungsleitfaden (NUWEL) 'long list' is to ensure that sustainability-related indicators are being considered by valuers but equally to avoid double accounting as some characteristics such as location may in fact be counted twice if the sustainability indicators are being looked at separately.

However, what constitutes value for the individual market participant still very much depends on their respective objectives and motivations. In the first instance, financial yields more or less capture the traditional concept of value in commercial real estate. From an investment angle, these financial yields usually are in the shape of direct monetary gains upon sale or rental but can also take on a risk management dimension as quicker turnaround times and reduced void periods also directly translate into financial value, as does continued insurability due to reduced physical risks to the asset such as flooding, subsidence, etc.

In other words, if the sustainable refurbishment of a building contributes to either of the above, for an investor there would be value in carrying out the refurbishment, especially in markets that favour sustainable buildings over conventional ones where a 'business as usual' approach could lead to the building becoming difficult to sell or rent or ultimately completely obsolescent. It is worth noting that in such more mature markets the value increments for sustainable buildings upon sale or rental eventually tend to level out, as being 'sustainable' is no longer an additional set of features but becomes the market standard. In *Dynamics of Green Building* (Eichholtz *et al.*, 2011), their second 'green premium study' on building clusters in the US commercial market, Eichholtz *et al.* found that the premiums more sustainable buildings commanded in 2010 were slightly lower than in their 2007 study (Eichholtz *et al.*, 2009a) which they attributed to the increased number of such buildings in the market. The same phenomenon was noted by Salvi and Syz (2010) with regard to the value of the Swiss Minergie label. Therefore, in the medium to long term it is expected that current 'green' price premiums for sustainable buildings will turn into 'brown' price discounts for the conventional existing building stock. Whereas lower operating costs, such as lower energy, gas or water bills may be of lesser direct importance to investors, they are certainly all the more important for occupiers and can thus positively influence rental value and overall rentability.

For those investors who are taking a more long-term view, sustainable refurbishment investments can also mean 'future-proofing' portfolios against an inevitable further tightening of regulatory frameworks.

A slightly more intangible reason to opt for sustainable refurbishment but one that is increasingly gaining traction both in empirical research and in discussions amongst market stakeholders is that sustainable buildings have the added potential of possibly leading to improved workplace productivity and employee health, i.e. higher output, fewer sick days, better employee retention resulting in bottom line benefits. A previous section showed significant existing gaps regarding the data landscape for the more 'traditional' value-determining aspects of sustainability, such as environmental performance. This is definitely even more so the case with data relating to the relationship between workplace productivity and health and sustainable buildings. It is extremely difficult to objectively measure productivity as it is also linked to other influencing factors such as management style and the overall organisational culture which means that the sustainable building productivity and employee satisfaction aspects cannot so easily be isolated. Consequently, workplace productivity is not a standard item within financial metrics yet as more research is needed to underpin present findings which are currently mainly from studies covering schools and hospitals.

In conclusion, from an investment point of view there are several compelling risk management elements that would underpin the business case of investing in sustainable refurbishment of commercial buildings:

- reducing potential risks to future income flow, depreciation and liquidity (risk of obsolescence);
- reducing risks to future funding and financing;
- reducing risks with regard to changing occupier behaviour;
- reducing risks resulting from a future legislative environment.

Having examined the specific reasons why investors might prefer to have sustainable real estate portfolios, for the business case rationale it is equally important to understand what makes occupiers want to rent these buildings. One might expect that from an occupier perspective, the aforesaid lower operating costs would be the key driver for opting for a sustainable rather than a conventional building. But that is only half the story. The 2009 RICS research report *Why Do Companies Rent Green* (Eichholtz *et al.*, 2009b) concludes that there are certain groups of occupiers for whom these operational savings are superseded by other strategic considerations.

In an age where it can take years to build up consumer trust and a positive brand image but only seconds to destroy it due to greater public scrutiny and the almost virulent speed at which news travels, for certain occupiers building and maintaining corporate reputation is in a value class of its own. It is also an age of greater accountability, in which organisations endeavour to show that they are committed to operating a responsible business model. Organisations often struggle to successfully communicate their Corporate Social Responsibility (CSR) and/or Sustainability programmes to their employees, customers and stakeholders. Occupying a sustainable building or individual offices can thus be the physical embodiment of these programmes or become part of the marketing for a retailer.

In terms of their specific profile, Eichholtz *et al.* divide organisations that are most likely to value sustainable buildings into the following clusters:

- organisations that compete for highly skilled human capital, such as law firms and financial institutions;
- government agencies and non-government organisations (NGOs);
- organisations that are exposed to higher reputational risk due to their overall business focus, such as mining firms, oil or tobacco companies, etc.

Figure 2.3 illustrates in a simplified way how corporate decision-making to undertake sustainable refurbishment of a building is quite complex and by no means one-dimensional. Different drivers to undertake sustainable refurbishments may be at play depending on the market participant's individual profile and strategic business objectives. The question of which

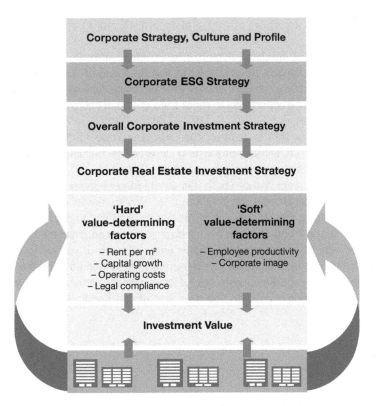

Figure 2.3 Corporate decision-making drivers for sustainable refurbishment investments in commercial property

Source: RICS

sustainability-building characteristics influence investment decision-making is thus directly dependent on the overall corporate strategy, culture and profile, and Environmental, Social and Governance (ESG) mission which in turn sets the parameters for the organisation's investment and real estate investment strategy. For an organisation with a strong ESG commitment, sustainable real estate characteristics have the potential to positively influence the organisation's investment value both as so-called 'hard' and 'soft' value-determining factors.[7]

All this presents a major challenge to valuers as what constitutes 'Market Value' is becoming increasingly complex and multi-layered. Integrating sustainability considerations into valuation is also an issue of creating greater transparency: clients need to understand the valuer's thought processes but valuers equally also need to better understand their clients' specific motivation. Or as expressed by a practicing valuer: what counts is not only the valuation result (i.e. the single-point value estimate) but the advisory services, i.e. information and explanations that come with the valuation.[8]

FINAL WORDS

As this chapter has illustrated, developing, strengthening and communicating the link between sustainable refurbishments and the value of buildings has an important part to play in the EU's energy efficiency and renewable energy strategy, offering a huge opportunity to meet the EU's 2020 energy efficiency and renewables targets.[9]

However, for this to happen the business case for wide-scale investment in the refurbishment of existing commercial buildings and the crucial role the valuation process plays in this must be made, explained, disseminated and incorporated into both real estate market decision-making.

Given the increasingly complex nature of how sustainability impacts a commercial building's asset value, it is absolutely essential to put mechanisms in place that will both improve overall data collection and management in Europe as well as widen the knowledge and skill base of valuers with regard to sustainable refurbishment investment options when advising their clients regarding the potential benefits of investments into sustainable refurbishment, resulting in an improved data and information flow along the whole value chain.

NOTES

1   These workshops were held in the framework of the Intelligent Energy Europe-funded project TRAINREBUILD.
2   The Royal Institution of Chartered Surveyors (RICS) first issued an Information Paper on Sustainability and Commercially Valuation in 2009 (with a subsequent updated version to be published in the autumn of 2013) followed by a residential version in 2011.
3   The Sustainable Building Alliance is an international non-profit organisation bringing together operators of building rating tools and certification, standard setting organisations, national building research centres and key property industry stakeholders. The objective of the Alliance is accelerating the adoption of sustainable building practices through the promotion of shared indicators for building performance assessment and rating.
4   Property constitutes one of the work streams within the United Nations Environment Finance Initiative. The purpose of the Property Working Group (PWG) is to encourage property investment and management practices that achieve the best possible environmental, social and financial results.
5   The United Nations-supported Principles for Responsible Investment (PRI) Initiative is an international network of investors working together to put the six Principles for Responsible Investment into practice.
6   The Institutional Investors Group on Climate Change (IIGCC) is a forum for collaboration on climate change for European investors.
7   The reality of corporate decision-making obviously tends to be much more complex as a number of considerations such as site specifics, location and individual vested interests of corporate departments may also be at play.
8   Personal correspondence with Professor David Lorenz, Karlsruhe Institute of Technology, Germany.
9   The EU's 2020 targets for energy efficiency and renewables are part of the overall EU climate and energy package which is a set of binding legislation aiming to ensure the European Union meets its ambitious climate and energy targets for 2020.

REFERENCES

Bio Intelligence Service, Lyons, R. and IEEP (2013) *Energy Performance Certificates in Buildings and Their Impact on Transaction Prices and Rents in Selected EU Countries*, Final report prepared for European Commission (DG Energy).

Brounen, D. and Kok, N. (2010) *On the Economics of Energy Labels in the EU Housing Market*, London: RICS.

Buildings Performance Institute Europe (BPIE) (2013) *A Guide to Developing Strategies for Building Energy Renovation*, Delivering Article 4 of the Energy Efficiency Directive.

Chegut, A., Eichholtz, P. and Kok, N. (2012) *Supply, Demand and the Value of Green Buildings*, London: RICS.

Eichholtz, P., Kok, N. and Quigley, J. (2009a) *Doing Well by Doing Good? An Analysis of the Financial Performance of Green Office Buildings in the USA*, London: RICS.

Eichholtz, P., Kok, N. and Quigley, J. (2009b) *Why Do Companies Rent Green?* Real Property and Corporate Social Responsibility, London: RICS.

Eichholtz, P., Kok, N. and Quigley, J. (2011) *Sustainability and the Dynamics of Green Building: New Evidence on the Financial Performance of Green Buildings in the USA*, London: RICS.

European Commission (2010) Directive 2010/31/EU on the *Energy Performance of Buildings*, Official Journal of the European Union, L 153/22.

Institutional Investors Group on Climate Change (IIGCC) (2013) *Protecting Value in Real Estate: Managing Investment Risks from Climate Change*, Brussels: IIGCC.

Kok, N. and Jennen, M. (2012) 'On the value of energy labels in the European Market', in *Energy Policy*, Amsterdam: Elsevier.

Meins, E., Lützkendorf, T., Lorenz,D., Leopoldsberger, G., Ok Kyu Frank, S., Burkhard, H.-P., Stoy, C. and Bienert, S. (2011) *Nachhaltigkeit und Wertermittlung von Immobilien, Leitfaden für Deutschland, Österreich und die Schweiz (NUWEL)*, Zurich: Center for Corporate Responsibility and Sustainability (CCRS), University of Zurich.

RICS (2009) *Energy Performance and Value Project*, Interim report from RICS to Department of Communities and Local Government, London: RICS.

Sahadi, B. *et al.* (2013) *Home Energy Efficiency and Mortgage Risks*, Washington, DC: The Institute for Market Transformation.

Salvi, M. and Syz, J. (2010) *Der Minergie-Boom unter der Lupe*, Zurich: Center for Corporate Responsibility and Sustainability, University of Zurich.

United Nations Environment Programme Finance Initiative (2011) *An Investor's Perspective on Environmental Metrics*, Geneva: UNEP FI.

Van Eck, A. (2008) *De 'Willingness to pay' voor een energiezuinige nieuwbouw wooning*, Delft: TU Delft.

Wameling, T. (2010) *Immobilienwert und Energiebedarf: Einfluss energetischer Beschaffenheiten auf Verkehrswerte von Immobilien*, Stuttgart: Fraunhofer IRB Verlag.

World Green Building Council (WGBC) (2013) *The Business Case for Green Building: A Review of the Costs and Benefits for Developers, Investors and Occupants*, WGBC.

3
RESILIENCE TO A CHANGING CLIMATE
Bill Gething

Climate change sets two inextricably intertwined agendas for designers and their clients: Mitigation (reducing the emissions that drive climate change in order to limit the magnitude of that change) and Adaptation (designing differently to take proper account of the associated changes underway). The more successful we are with the former, the less will be the need for the latter.

Some design strategies can tackle both agendas simultaneously but some aimed at one work in direct opposition or have unforeseen consequences for the other. For example, where an existing building is insulated and made more airtight in order to reduce winter heating energy consumption, overheating can occur in summer if purge ventilation is not considered properly. The implications for both agendas need to be carefully followed through for any strategy.

Rising energy prices and growing legislation, reflecting global efforts to reduce carbon emissions, have made energy and carbon reduction key drivers of any refurbishment programme and are increasingly well understood by clients and the construction industry. However, we have been slower to register the reality that our climate is, indeed, changing and that design parameters must acknowledge those changes so that our buildings can cope effectively with environmental conditions that may be different from those in the past. Some change is inevitable: we can do nothing about the changes that have already occurred and, even if we cut emissions immediately to zero, the momentum in the climate system would continue to drive change until the mid-century.

The variety in the form, character and detailing of buildings and their urban context around the globe is testament to the fundamental influence of local climate on the evolution of their design. Clearly, a change in that climate implies a corresponding re-evaluation of established design custom and practice and offers the potential for intelligent innovation to enable our buildings to respond to changing environmental circumstances.

Architects, as the principal controllers of building form and the integrators of input from all members of the design team, are in a key position to lead this re-evaluation. As professionals, it could be argued that they also have a responsibility to make their clients aware of the need to consider potential impacts of climate change on a project.

Unfortunately, whereas there are regulations and standards that set the framework for tackling the mitigation agenda, there are, as yet, few if any agreed standards on which clients, developers and designers can base a 'reasonable' approach to adaptation. The issue cannot be ignored, however, and there is already a somewhat disturbing interest from the legal profession who scent a potentially lucrative future workload pursuing the designers of buildings that have not adequately allowed for climate change. The view being that 'everyone knows about climate change' so how could a designer reasonably ignore the issue?

In the absence of guidance and standards, decisions on the development of an adaptation strategy must, therefore, be made on a project by project basis, taking into account the particular vulnerabilities of the building type, location and its intended use.

HEADLINE IMPACTS

Climate projections are produced by a number of organisations from around the world using large computer models of the interaction between climate system and a range of different greenhouse gas emission 'pathways' or scenarios that have been developed by the IPCC (Intergovernmental Panel on Climate Change) to reflect different demographic, social, economic and technological changes.

An overall picture is provided by the IPCC Assessment Reports on global climate change produced by IPCC since 1990. The most recent complete report is the Fourth, produced in 2007, based on outputs from twenty-three climate models, and elements of the Fifth Report, which will be completed in 2014, are now starting to emerge (IPCC, 2007).

The availability, scope and content of future climate data inevitably varies between different countries. In the UK, for example, a comprehensive set of information was produced by the Met Office Hadley Centre in 2009, known as the UKCP09 projections (DEFRA, 2010). Information provided for twenty-six atmospheric variables is available on a 25 km^2 grid across the country up to the end of the century, using three selected IPCC emissions pathways: B1, A2 and A1F1, denoted Low, Medium and High for ease of reference. Values are provided relative to baseline data for the period 1960–91 (useful for understanding relative changes) and absolute values (more useful for definitive analysis). Tailored data is available in the form of tables, maps and graphs for the three emissions pathways for time periods throughout the century via a freely available User Interface.[1] UKCP09 also includes 'Weather Generator' to provide even finer grained information; however, this requires considerable computer power and expertise to use and is not suitable for use in general design practice.

The wealth of detailed information may appear daunting and in order to understand the principal issues which must be tackled, a general picture is perhaps more helpful. Taking the UK as an example, this can be summarised as follows:

- warmer and wetter winters;
- hotter and drier summers;
- an increase in extreme events;
- rising sea levels.

Temperatures are projected to rise, more so in the south than the north, and whereas relatively little change is projected in total annual rainfall, the seasonal pattern is likely to be different with more in winter and less in summer.

In similarly general terms, impacts on the built environment can be considered under three simple headings:

- comfort and energy use – particularly in increasing likelihood of overheating in summer;
- construction – changes in the behaviour of materials, impacts on detailing and foundation design for shrinkable soils;
- water – too little (the impact of changing rainfall patterns on water supply) and too much (flooding from a variety of sources).

These are discussed in later sections with charts that further illustrate the kinds of design issues that might be considered under these headings, cross-referenced to a timescale context for decision-making.

CONTEXT AND TACTICS

It is important to note that impacts will vary geographically, both in terms of broad regional differences and the specific circumstances of a particular location. Overheating, for example will be more of an issue in warmer regions and in urban areas subject to the heat island effect, whereas flooding may be the key design driver for sites close to rivers or beside the sea but much less of a concern inland on higher ground.

As with the mitigation agenda, design discussion tends to focus on new build; however, converting the existing stock to suit changing climate is really the key issue. This is particularly relevant in the case of adaptation as climate change is a moving target with considerable uncertainties about the rate and magnitude of change (although not the direction of travel). It is therefore not economically or practically feasible, even for new build, to 'climate-proof' against all eventualities. The aim should be to develop a strategy for ongoing interventions with an understanding of when these might need to be implemented and recognising that this may change. Clearly some options are available for new build, such as resilient foundation design, that are less applicable to retrofit; however, upgrading facades or replacing mechanical services are opportunities to incorporate adaptation measures that can be exploited when a building is retrofitted at intervals throughout its lifespan.

The inextricable link between mitigation and adaptation means that basing an adaptation strategy on an increased use of fossil-fuelled energy is not a viable long-term option – a strategy that would be perhaps the clearest example of attempting to deal with one agenda at the expense of the other. It is also difficult to imagine that the cost of energy will fall in relative terms in future and it is far more likely that it will become increasingly expensive for the foreseeable future and perhaps in shorter supply. Under these circumstances, following the usual energy hierarchy makes complete sense:

- first, to exploit design skill to maximise the passive potential inherent in an existing building;
- second, to use servicing systems that are as efficient as possible to temper the internal environment using as little energy as possible;
- third, to exploit renewable resources to deliver that energy as far as possible.

It is also worth bearing in mind that design parameters may change over time and that there may be options to explore with a client either to redefine comfort criteria more flexibly or to recognise that behavioural adaptation might provide an additional level of resilience. It is difficult to predict how conventions will change and how much we will adapt in the context of future change. For example, although not generally an option when considering an individual refurbishment project under current circumstances, changing working hours could significantly reduce cooling requirements, as could relaxing dress codes, etc. (see Chapter 7).

Similarly, although all potential impacts of climate change should be assessed for any project, it should not be assumed that all issues should be solved at the individual building level. Some aspects of adaptation may be better dealt with at a larger scale and, while this may not be an option for many projects, it is worth considering alternative options when circumstances arise. For example, where a building is part of a larger campus, an initiative to reduce mains water consumption by rainwater harvesting is likely to be better value for money, both in terms of capital cost and ongoing maintenance, at a campus rather than individual building level.

Which climate to design for?

Whereas climate projections for a given variable for a given time period and emissions pathway can be based on a single climate model, a more statistically robust approach is to combine results from a number of models and present these as a probabilistic range rather than a single value. This is the approach taken in the UK for the UKCP09 projections which uses the same language as the IPCC to ascribe 'likelihood' to particular statistical values based on how well the models correlate with each other and with historical measurements when they are back-cast; the most commonly used being the 'central estimate' (50th percentile), the 'likely' range (33rd to 66th percentile) and the 'very likely' range (10th to 90th percentile).

The approach may be statistically robust but adds complexity in a field that is already rich in uncertainties. CIBSE (the Chartered Institution of Building Services Engineers) and UKCIP (the UK Climate Impacts Programme) have developed the Probabilistic Climate Profile (ProCliP) methodology to produce a set of graphs illustrating the range of projected values for different variables for a number of locations to help designers select appropriate design parameters for their project (see Figure 3.1). Although CIBSE only produces graphs for UK locations, the methodology is applicable for probabilistic projections anywhere.

When considering overheating, the range of values for summer mean daily maximum temperatures gives a useful indication of changing conditions. The ProCliP graph below shows the range of projected values through the century for London, indicating the central estimate and the 'likely' and 'very likely' range for each of the UKCP09 emissions pathways.

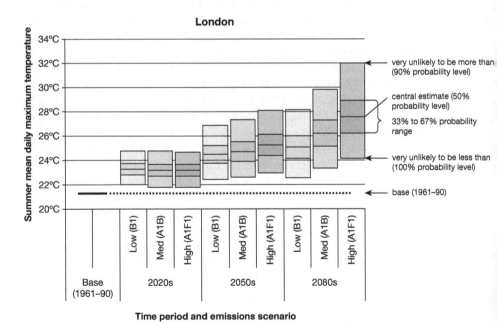

Figure 3.1 ProCliP graph showing the summer mean daily maximum temperature

Source: CIBSE (Chartered Institution of Building Services Engineers)

One of the first points to note about the graph is the difference between the dotted base case value (1961–90) and the spread of values for the 2020s. In a changing climate, designed criteria based on averaging historic data will inevitably lag behind current conditions and there is a strong case for using 2020s projected figures as an absolute minimum. After all, the 2020s figures represent an average over the thirty-year period centred on 2025, i.e. 2010–40. We are already in the '2020s'.

The graph illustrates how little difference there is between the scenarios early on in the century, the difference increasing as the century moves on. It also highlights that, at any point in the century, the difference between the same percentile values for the different emissions pathways is less than the probabilistic range for an individual emissions pathway. The choice of where a project should sit on the probabilistic range, perhaps to reflect a particular vulnerability of a project to a given environmental condition, thus has more impact choosing between emissions pathways. A central estimate might be a reasonable starting point, but for a building in an area that is potentially prone to flooding, for example, it would make sense to use a higher percentile for, say, peak winter rainfall than would be necessary for a building in a less sensitive location.

A building's vulnerability to higher temperatures, for example, may be significantly different to its vulnerability to changes in rainfall patterns and it may therefore be appropriate to select different combinations of probabilities, time periods and pathways to test different aspects of the design.

This graph also illustrates the moving target of climate change. Even a value at the upper end of the probabilistic range early on in the century tends towards an average, central estimate for the middle of the century and becomes a value at the low end of the probabilistic range by the 2080s. Under these circumstances, it is perhaps easier to approach the problem from the other direction. Rather than try to establish a logic for choosing a given combination of probabilistic value, emissions pathway and time period, one can select a value representing, say, a two degree increase in summer mean daily temperature and review this in terms of how long it remains a credible choice through the century, given the particular circumstances of the project.

In establishing design strategies it is also worth investigating whether, whilst the change in conditions may be gradual, there are particular thresholds where one approach can no longer cope and an alternative strategy must be used which may have very different implications for the building. For example, the point at which it is no longer possible to maintain comfortable conditions without some mechanical cooling with all the implications for duct routes, floor-to-floor height, plant space, etc. that follow.

When first getting to grips with these issues, it can seem that the uncertainties are overwhelming and that steering a defensible path through them is an almost impossible task. It is perhaps reassuring to reflect on the extraordinarily significant changes that commercial buildings have been through in recent years which their designers could not possibly have foreseen, for example the proliferation of the personal computer, which has changed the heat balance of office buildings to as radical an extent as some of the changes we are now considering.

The decision to refurbish, rather than demolish and replace, a building demonstrates that its fundamental character, construction and configuration have stood the test of time, albeit that some elements may need replacing or upgrading to current standards and it may need reorganising to accommodate changes in the functions to support for the next phase in its life. That said, some impacts of climate change will, in due course, have a very significant bearing on whether or not to refurbish the building. There will come a time, for example, where the increasing risk of flooding reaches the point that there is no realistic alternative but to abandon areas of low-lying land (including long-established urban areas), let alone individual buildings.

Design data available

For some aspects of environmental design, the ProCliP approach can provide values that can be used directly, for example, for manual cooling calculations. However, generally basic climate data needs 'translating' into the kinds of data sets that are commonly used in design practice. Limited inroads have been made into this task; however, a notable exception is an extension to the range of standard weather files used in industry-standard dynamic simulation software packages to include future weather files to analyse the thermal behaviour of buildings. These are generally provided on a national basis such as those available for the UK from CIBSE or Exeter University (the free to download PROMETHEUS data sets); however, there are sources for any location worldwide, such as those available from Meteonorm.

COMFORT AND ENERGY USE

There are many opportunities and challenges for the designer when designing for comfort:

Keeping cool – internal spaces
Fabric

- optimise fabric insulation level;
- balancing daylight and solar gain;
- shading – building form and orientation;
- shading – manufactured – fixed;
- shading – planting;
- glass, glazing and film technologies;
- protection – layered/ventilated facades and roofs, green/brown/ 'blue' roofs;

- reflective materials;
- fabric energy storage plus purge ventilation – phase change materials;
- ceiling heights – air movement, stratification, ceiling fans;
- transpiration cooling – planting, water, green/'blue' roofs and walls.

Control

- user focused design – intuitive and robust controls, clear instructions;
- shading – manufactured – movable;
- purge ventilation – control, free area, noise, pollution, security, weather, bugs, etc.;
- night ventilation – control, free area, noise, pollution, security, weather, bugs, etc.;
- minimal summer ventilation strategy (where external air temperature is too high for effective purge vent);
- timetabling – working hours/siesta;
- changing perceptions of comfort in a changing climate (see Figure 3.2).

Systems

- minimise gains from equipment and lighting;
- mixed mode cooling strategies;
- low-energy/carbon cooling systems;
- integrated service design – efficient layout, space for ducts, etc., ease of access for (screwdriver-free) maintenance;
- installation, commissioning, explanation, maintenance and monitoring of unfamiliar/innovative systems;
- ground-coupled cooling – air-based – earth tubes, labyrinths;
- ground-coupled cooling – heat pump-assisted – slinkies, thermal piles;
- water source cooling – bore hole to aquifer, water body source.

Keeping cool – spaces around buildings

- built form – building to building shading;
- access to external space for relief;
- shade – planting – for people, spaces, buildings and vehicles; consider irrigation/drought resilience of species;
- shade – manufactured – potential interrelationship with renewables;
- large-scale green/'blue' space/roofs to mitigate urban heat island effect (micro and macro);
- exploit cooling wind effects.

Keeping warm

- consider extremes as well as trends in average temperatures;
- evaluate merits of reclaiming heat from ventilation;
- take account of change in balance between space heating and other loads (e.g. hot water).

Historically, the environmental design of buildings in temperate climates has tended to focus on keeping warm in winter and, more recently in response to the mitigation agenda, to improve the building fabric, with better insulation and improved airtightness, to reduce the energy needed to do so. Outside the heating season, we have expected buildings to operate passively, maintaining comfortable conditions even at the peak of summer by balancing gains from people, equipment, lighting and the sun, with losses by conduction through the fabric and ventilation. Traditionally, thermally massive construction has also helped smooth out diurnal fluctuations in temperature, using the building structure to absorb heat during the day and release it by purging the building with cool night air.

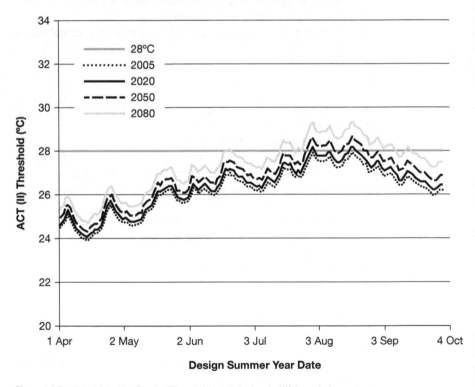

Figure 3.2 Predicted Adaptive Comfort Threshold levels for Leeds, UK through the century

Source: Arup UK

Figure 3.3 illustrates the elements of the balancing act. Fabric thermal storage or thermal mass is shown acting as a kind of shock absorber

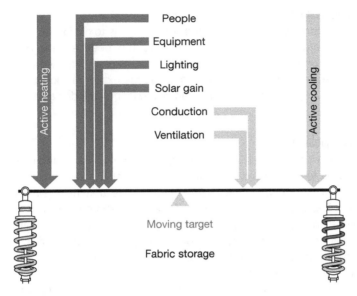

People

Equipment

Lighting

Solar gain

Conduction

Ventilation

Active heating

Active cooling

Moving target

Fabric storage

Figure 3.3 Balance of heat inputs and losses from a building in a cool climate

Source: Bill Gething

with each passing into and out of the mass to even out fluctuations. The aim in a temperate climate is to exploit the passive potential of a building so that it is in 'free running' mode, with heat gains and losses balanced for as much of the year as possible and adding mechanical heating or cooling to tip the balance when necessary to maintain comfortable internal conditions.

It is worth noting that, even if it is still being used for its original purpose, the energy balance that may have pertained when a building was constructed will almost certainly have changed significantly by the time it comes to be refurbished. It is a testament to the resilience of many existing buildings that they have been able to accommodate changes such as higher densities of occupation and equipment loads that could not been envisaged by their original designers.

Different levels of refurbishment offer greater or lesser oppor-tunities to affect all the elements of the balance to set the building up for the next phase of its life, from improving the efficiency of mechanical systems to altering the size and distribution of heat loads from people, equipment and lighting, and even to altering characteristics of the building envelope through changes to the facade or exposing previously inaccessible thermal mass.

A warming climate adds a level of complexity to the balancing act in that the pivot point effectively moves to the right through the lifetime of the building. Our design focus therefore needs to shift correspondingly to make sure that our efforts to improve wintertime performance are balanced against the need to avoid or minimise mechanical cooling. Valuable lessons can be learnt by looking at traditional solutions, codes and regulations used in warmer climates; however, it should be noted that these cannot necessarily be transferred wholesale. Some aspects of climate will not change, such as sun angle, and it is important to recognise that the challenge is to design for new combinations of environmental factors that have not been experienced before. A corollary of the phenomenon is that our perception of climatic zones is likely to change, particularly in transition zones, to the extent that locations which are currently regarded as temperate can no longer be regarded as such.

Understanding the problem

It follows that an essential step in developing an effective year-round low-energy strategy is to analyse and understand the heat balance for a building, breaking the analysis down as necessary into spaces that have similar characteristics. This is particularly important for a refurbishment which may be made up of elements of very different thermal characteristics. Even with a building that appears to be broadly homogeneous, there may be significant differences between spaces depending on orientation, over-shading and ventilation arrangements and even for fully open-plan buildings, the conditions at the perimeter are likely to be very different from spaces more remote from the facade.

Climate change tends to exploit weaknesses in the characteristics of an existing building that may already have become evident when it has been stressed by what have historically been relatively extreme conditions, as these will increasingly become the norm. One of the great advantages of working with an existing building, particularly if it is to be used for similar purposes after refurbishment, is that the design team have the luxury of a full-scale working model to help understand how it works. Simply asking occupants and facilities management staff how existing spaces behave through the day and through the seasons can be an extremely effective shortcut to identify problem areas and to act as a check on designers' preconceptions. Provided time and circumstances permit, it can be similarly invaluable to monitor pre-existing environmental conditions (see Chapter 11). This hard data can put the callipers on user perceptions and help validate theoretical modelling work at the outset of refurbishment design work. It is at this point that the team has the greatest opportunity to design its way out of problems rather than have to tackle them head on, for example, by exploiting the potential to replan functions into spaces with environmental characteristics that are better suited to the activity.

To develop and test robust adaptation strategies will require thermal modelling, typically dynamic simulation, but the investment will be repaid by having a clear understanding of what the 'problem' is, so that design effort can be focused on the right issues for each space or element type. This can take time, both to carry out then to analyse and then to iterate adaptation options and so it is essential that the analysis starts as early as possible in the design process, so that findings can provide direction to the evolution of design proposals rather than support a rearguard action after crucial decisions have been made.

The process also generates large amounts of data and it is important to establish robust systems to keep track of options and iterations, to streamline analysis and convey results in an understandable way. One useful technique is to produce a 'Heat Map' for the building. This uses simple Excel spreadsheets that list spaces, arranged, say, by facade orientation, with hours over temperature thresholds gathered from simulation plotted against them and using Excel's conditional formatting to highlight where comfort thresholds are exceeded. This gives a quick overview of the effect of changes in environmental conditions and highlights where in the building there might be particular issues.

Comfort is intimately linked to energy use in buildings and renovation must consider the two together to optimise sustainability. The main issues are to manage heat gains, adapt the building facades, provide adequate ventilation, optimise the use of thermal mass and mechanical services, as well as the addition of external features such as planting.

Reducing unnecessary gains

Lighting and equipment

Although incidental gains from equipment and artificial lighting may make a contribution to winter heating, this is, effectively, using (typically carbon intensive) electricity to heat the building rather than relatively carbon-efficient heating fuels such as gas. Installing the most energy-efficient equipment and controls will reduce or avoid the need for cooling as well as reducing carbon emissions and running costs. This is a win-win strategy tackling both the mitigation and adaptation agendas and is the key adaptation strategy for deep-plan space, where the impact of external conditions can be quite limited as compared with internal gains.

There may be circumstances where there is 'spare' heat from essential electrical equipment that could usefully be used in winter, for example, from server rooms, and exploiting this potential may inform space planning decisions as to whether to centralise or distribute these kinds of functions around the building. Again, striking an appropriate balance is essential, based on a clear understanding as to when the waste heat is a benefit and making sure that, when it is not, it does not add to cooling loads.

Good levels of daylight can clearly avoid the need for artificial lighting; however, achieving this effectively requires careful design and cannot be achieved by simply increasing the amount of glazing, even if this is an option for a refurbishment. Again, this is a balancing act, to provide useful daylight without unwanted glare or solar gain. For top floors, there may be an opportunity to incorporate north-facing roof lights (south-facing for southern latitudes) which can bring useful daylight deep into the plan without significant solar gain. Glazing to vertical facades needs more careful consideration however, to get maximum penetration of daylight into the plan without excessive glare for those at the perimeter, which will force them to lower blinds and cut out daylight to those deeper in the plan – the all-too-familiar blinds down, lights on syndrome. For buildings with natural ventilation via conventional windows, lowering a blind to control glare or solar gain will often also block ventilation, thereby, compounding the problem.

Controlling solar gains

For shallow-plan buildings and at the perimeter of deeper floor plates, solar gain is likely to be a significant heat input, raising internal air temperatures and warming occupants by direct radiation. In temperate climates these effects may be welcomed in winter, reducing energy consumption for heating, but may overload the space outside the heating season (which can be quite short in a well-insulated, airtight and densely occupied commercial building). The control of solar gain through the seasons and through the day is therefore a key strategy in maintaining a comfortable heat balance for these spaces.

Facade refurbishment or replacement is a key opportunity to reset the balance between heat losses and gains and to introduce more active solar control measures to give the building enhanced adaptation capacity. This opportunity is likely to occur at regular intervals in a building's life and can play a key part in ongoing adaptation strategies that can be accelerated or slowed down as the magnitude and rate of climate change becomes clearer (see Chapter 9).

Fortunately in temperate zones the difference between low winter and higher summer sun angles offers the potential to incorporate overhangs or horizontal shading devices to southern elevations (northern in southern latitudes) to provide useful daylight whilst allowing direct sun to provide useful warmth in winter but excluding it in summer.

Eastern and western facades are more problematic as sun angles are lower and as a result these facades can be the most heavily loaded in terms of solar radiation. Whilst significant benefits can be gained by incorporating vertical 'blinker' shading devices when replacing windows or glazed elements, perhaps in combination with horizontal shading components, it may be necessary to incorporate movable shading devices for effective control.

External fabric

Refurbishment offers the opportunity to upgrade the solid external fabric of the building in order to reduce year-round energy consumption and carbon emissions. This can be achieved, depending on the nature of the building concerned and the extent of intervention, by internal or external insulation or wholesale replacement of the facade. The merits and caveats associated with these approaches are discussed in Chapter 8; however, as a general rule, the principal driver for establishing an appropriate level of thermal performance for solid building elements is to reduce winter heating demand and the resulting quality of fabric is generally more than adequate to ensure that unwanted summertime gains through the fabric are negligible.

Natural ventilation, airtightness and thermal storage

In response to the mitigation agenda, much design effort has focused on making our buildings more airtight and developing methods to provide minimum acceptable ventilation in winter, often supplemented with systems to reclaim heat lost in the process. We have tended to focus less on providing sufficient, controllable natural ventilation, arguably the key low-energy strategy to maintain comfortable conditions outside the heating season. Where natural ventilation and cooling are an option, often in cooler climates and smaller buildings, ventilation and fabric thermal storage provided by thermally massive construction are also inextricably linked. Thermal mass does not provide cooling per se. All it can do is dampen the thermal behaviour of a building by evening out high and low temperatures, typically on a diurnal basis. Without adequate ventilation to purge heat absorbed in the day, heat will continue to build up in the fabric during a hot period, potentially exacerbating overheating rather than reducing it. With effective night-time ventilation (that also needs to be secure, rain-, vermin- and bug-resistant) thermal mass is an asset, absorbing heat in the day and releasing it when the space is purged by cool night air, transferring the benefit of relatively low night-time temperatures typical of temperate climates, to daytime occupation. High levels of daytime ventilation are likely to be effective for much of the year and the same requirements for security, etc. apply. In noisy environments, attenuation may be necessary and accommodating this in a refurbished facade may be a significant technical and aesthetic challenge.

With a refurbishment there are generally limited opportunities to alter the fabric storage capacity of a building, although there may be opportunities to expose existing thermal mass by removing suspended ceilings, for example, which may also provide loftier spaces allowing more stratification of hot air above the occupied zone. Care should, of course, be taken not to unthinkingly throw away the potential benefit of existing thermal mass by adopting an inappropriate internal insulation strategy. Phase change materials offer the opportunity to add fabric thermal storage in refurbishments without the weight associated with conventional thermally massive construction. A number of dynamic simulation tools are now able to model their effect.

Natural ventilation also provides a cooling effect for occupants through air movement even when external temperatures are high, but again this needs to be carefully controlled so as to avoid disturbance. Ceiling fans are a low-energy means of providing air movement mechanically provided the space is high enough to accommodate them.

Mechanical ventilation and cooling

In a warming climate, heating systems need to be designed for current environmental conditions, whereas cooling requirements need to be based on an assessment of likely conditions by the time the building is next refurbished or services replaced. For buildings that can operate purely under natural ventilation at present, it is wise to investigate the conditions under which they would need mechanical assistance and establish the likelihood as to whether they will cross that threshold in the lifetime of the current refurbishment. Depending on the likelihood of these conditions occurring, it may be prudent to explore the physical implications of having to accommodate plant, equipment, service runs, ductwork routes, etc. on the current refurbishment proposals in case there are fundamental, longer term implications for the adaptability of the building.

Mechanical ventilation and cooling systems will frequently be needed in commercial buildings to provide adequate cooling at least in summer. These should be provided as efficiently as possible, using strategies that allow natural ventilation to be exploited for as much of the time as possible. Displacement ventilation systems and the use of chilled beams may be suitable for refurbishment and ground-sourced cooling systems are used in some of the case studies. The design of mechanical services and their control is central to both comfort and energy use (see Chapter 8).

Green roofs and facades

The potential to introduce green roofs and facades is relatively limited for refurbishment work due to the difficulty of accommodating the associated loads or water infrastructure. Cooling effects related to transpiration are extremely difficult to model but there is little evidence that they are particularly effective in temperate climates. Green facades, at their simplest, growing plants up a wall, undoubtedly have a positive shading effect on the wall fabric which will reduce surface temperatures, reducing conduction gains but also beneficially reducing the amount of heat radiated back into external spaces.

CONSTRUCTION

There are several design opportunities and challenges related to the construction of the building:

Structural stability below ground

- foundation/underpinning design – subsidence/heave;
- retaining structures – changes in soils behaviour;
- underground structures in relation to water table – buoyancy, contamination, pressure;
- slope stability – new and existing.

Structural stability above ground

- lateral stability – wind (completed buildings and during construction);
- loading from ponding (resulting from overwhelmed roof drainage).

Fixings, materials and weatherproofing

- increased deflections of cladding systems – wind;
- fixing standards and deflections – wind and rain;
- detailing approaches to deal with increased exposure rating;
- consider changes in UV on materials;
- extended/extreme wetting – permeability, rotting, weight;
- extended/extreme heat – drying out, shrinkage, expansion, softening, etc.;
- extreme wind – effect on airtightness and detailing of airtightness layers.

Construction — work on-site

- temperature limitations for construction processes;
- changing definition of inclement weather;
- working conditions – site accommodation;
- working conditions – internal conditions of buildings under construction.

There are two facets of climate change that need to be taken into account when considering potential impacts on building materials, detailing and structural design. The first is the effect of changing general patterns of weather, particularly changes in total annual rainfall or its seasonal pattern, for which we have reasonably good data. The second is change in the magnitude and pattern of storms for which there is less consensus in the projections produced by climate models, one of the reasons being that the variability in these phenomena appears to relate more strongly to the varying position of storm tracks rather than any underlining climate change signal.

As yet, information on the future behaviour of materials is almost non-existent; however, designers can usefully carry out a pragmatic review of potential changes in behaviour by considering how these materials behave currently in different locations. Taking brickwork as an example: this is a porous material that relies for its weatherproofing function on an ongoing cycle of wetting and drying (from outside and inside) rather than presenting an impervious barrier to the elements. Wall thicknesses and construction methodologies have developed over time, which provide adequate performance, however these typically vary to reflect patterns of driving rain associated with geographical location and the degree of exposure of the site or building. Full fill cavity wall insulation, for example, has been found to be problematic above a certain level of exposure.

A change in seasonal patterns of precipitation will clearly have a corresponding impact on the balance of wetting and drying and thus the weather resistance of the wall. If the wall in question is also thermally upgraded in order to reduce energy consumption, the balance of wetting and drying can be further affected to the extent that there may be significant risks for the structure. For example, if internal insulation is applied to a solid brick wall, the amount of heat leaking from the interior to help dry out the brickwork will be reduced as well as dropping its temperature and, depending on the method of insulation, the ability of the wall to dry out to the interior will be reduced. A careful analysis needs to be undertaken to assess the combined effect of thermal upgrading and climate change, particularly if the wall contains organic components such as timber wall plates or lintels, etc.

Changing rainfall patterns are likely to lead to corresponding changes in soil moisture content which may have an adverse effect on shallow foundations in certain soil types, particularly shrinkable clays. This is more of a problem for domestic-scale buildings rather than larger commercial buildings which will normally have deeper foundations; however, when refurbishing older, smaller-scale buildings in areas of shrinkable clay, it would be wise to check on foundation depths.

As mentioned above, design data for future extreme wind patterns is limited and generally structural design standards on which engineers rely have not yet been updated to reflect the impact of projected changes. Generally there are considerable factors of safety incorporated in current engineering standards as far as wind loadings are concerned; however, designers would be wise to raise the issue with their clients and agree whether or not to design to augmented standards.

The allocation of responsibility for the detailed design of external wall assemblies as far as their resistance to extreme weather conditions is concerned is likely to be separate from that for structural design. In addition, the factors of safety incorporated in design standards and codes of practice for these aspects of design are generally lower than for structural design, and as a result, potential failure is more likely. Until standards are updated to include an agreed approach to accommodating climate change, decisions on whether to design any modifications to a facade to a higher standard than normal (for example, using an enhanced level of exposure for design calculations or testing) must be clarified on a project-by-project basis. An initial qualitative assessment of potential risks may be useful to scope the issues, for example, using a systematic 'traffic light' review of external wall options used by Arup on a project for Sheffield University in the UK.

Care should also be taken not to inadvertently take on responsibilities for the performance of existing facade elements under future climatic conditions.

WATER

There are several opportunities and challenges for the designer when considering water:

Water supply/conservation

- limitations on development in water-stressed locations;
- low water use fittings;
- grey water harvesting;
- rainwater harvesting;
- water bodies for irrigation/water storage;
- avoidance of water-intensive construction methods;
- alternatives to water-based drainage.

Flood – avoidance

- avoid development in flood prone locations (allowing for climate change);
- increased capacity or attenuation of roof and surface water drainage;
- risks from overwhelmed existing drainage (surface and combined);
- failsafe overflow routes for roof, surface water drainage and water/Sustainable Urban Drainage Systems (SUDS) features;
- permeable surfaces – exploit but consider impact of water-logging.

Flood – resistance/resilience

- flood defence – permanent;
- flood defence – temporary – external or at building perimeter;
- locate critical or vulnerable equipment or functions above flood level;
- evacuation routes and strategies;
- protected infrastructure (access, utilities, etc.);
- flood-tolerant construction and materials for rapid recovery;
- post-flood recovery strategies.

Landscape

- plant selection – consider resistance to waterlogging and drought;
- consider changes in ecology – flora, fauna, pests and diseases;
- irrigation – minimise and use efficient techniques;
- water features – benefits and challenges (e.g. mosquitoes);

- role of landscape, vegetation and hard surface on microclimate and heat island effect.

Flooding

As precipitation patterns change and sea levels rise, sites that are already at risk of flooding from natural features, such as rivers or the sea, may be exposed to greater risk and in some circumstances it is an unpalatable reality that an area's increasing vulnerability to flooding may ultimately be the fundamental reason for deciding to relocate rather than refurbish. Establishing the level of future risk, including the assessment of any existing flood defences, is therefore a key aspect in evaluating the feasibility of a refurbishment project.

Many of the measures to improve the resistance of a building to flooding that can be incorporated in new build projects (for example, raising ground floor levels or incorporating remote flood defences) are not generally available for refurbishment projects. Measures, such as temporary flood barriers at doorways, etc. in combination with non-return drainage valves and waterproofing walls up to a potential flood level, can increase resistance to short duration floods. However, these need to be very carefully coordinated to make sure that there are no weak points in the strategy where floodwater could bypass the defence regime. These approaches also have their limitations as deeper floods put lateral loads on walls which they were not designed to resist and involve hydraulic pressures which render many walling materials porous.

An alternative approach is to improve a building's resilience to a flood event so that essential services are unaffected and damage is limited, allowing finishes to be reinstated and the building brought back into service relatively rapidly and cheaply. Strategies would include locating essential plant, wiring and electrical equipment above potential flood levels and using construction methods and finishes that are resistant to water damage so that they can be cleaned down and dried out rapidly. Again, the strategy needs to be thoroughly thought through to avoid any weak spots.

In devising strategies to improve resistance or resilience the investment needs to be balanced against the wider picture. Whilst it may be technically possible to make a significant difference to an individual building, this may be rendered irrelevant if its workforce is unable to reach it or local services or transport infrastructure is disrupted for an extended period.

Flooding from rivers or the sea is generally well understood, however there are other potential flood risks to existing buildings. Changing patterns of extraction and seasonal rainfall may lead to raised groundwater levels affecting basements and the capacity of soakaways and sustainable urban drainage systems to dispose of surface water. Increased areas of impermeable surface (including buildings) in urban areas in combination with increasingly intense rainfall events may put strain on the adequacy of (potentially ageing) existing surface water drainage to cope with intense rainfall events (upstream and downstream of the site) which can lead to flooding. These potential risks need to be considered, albeit that the data on the behaviour of existing surface water drainage infrastructure under extreme conditions is generally less comprehensive than for marine or fluvial flooding.

Gutter and rainwater disposal design

For some buildings, risks from marine or fluvial flooding may be non-existent; however, all buildings need to take into account changing rainfall patterns in the assessment of the adequacy of their rainwater disposal systems, from the roof down to the point at which the water leaves the site. This is particularly relevant for buildings where overtopping of a gutter system would cause significant disruption or damage or where inadequate outlets could lead to unacceptable structural loads from backed up rainfall. Unfortunately, climate models typically provide rainfall data on an hourly or daily basis rather than at the short (minute level) timescale on which rainwater disposal design is based. Until standards are updated to include agreed allowances for climate change, design teams have little alternative but to make their own assessment and seek their clients' agreement on suitably robust design criteria.

Water conservation

Whilst the pattern of change in rainfall is not consistent around the globe, there are few locations where the availability of water is not already an increasing concern whether this is due to an inherent shortage of precipitation, competing demands for water, growing demand or concerns about the energy required to collect and distribute it. All of these factors, often coupled with regulation, have driven innovation in developing low water use fittings and appliances which are now widely available. It clearly makes sense to use the most efficient fittings that are economically available in any refurbishment both to reduce water consumption and where relevant, the energy needed to heat it.

At the scale of an individual refurbishment, although the circumstances may inevitably be more constrained than for a new build project, there may be options for substituting high-quality potable water with less highly treated, site-acquired water sources for uses such as toilet flushing or irrigation. These include the use of boreholes, rainwater harvesting or grey water recycling, although a realistic assessment of the ongoing maintenance requirements and costs should be made in addition to the capital costs and practicalities of providing these in the first place.

CONCLUSIONS

The designer needs to integrate the requirements of reducing energy use and CO_2 emissions with ensuring that the refurbished building is resilient to projected changes in the climate. Whilst energy conservation strategies and implementation are well understood and guidance readily available, the uncertainties and regional variations in climate change require careful analysis and the design of flexible strategies appropriate to local situation and building use.

NOTE

1   http://ukclimateprojections.defra.gov.uk/ (accessed 1 December 2013).

REFERENCES

DEFRA (2010) *UK Climate Projections, 2009: Briefing Report* (Geoff Jenkins, James Murphy, David Sexton, Jason Lowe, Met Office Hadley Centre, Phil Jones, Climatic Research Unit, University of East Anglia, Chris Kilsby, University of Newcastle), Version 2.

IPCC (2007) 'Summary for policymakers', in S. Solomon, D. Qin, M. Manning, Z. Chen, M. Marquis, K. B. Averyt, M. Tignor and H. L. Miller (eds) *Climate Change 2007: The Physical Science Basis*, Contribution of Working Group I to the Fourth Assessment Report of the Intergovernmental Panel on Climate Change, Cambridge and New York: Cambridge University Press.

4
MAKING SUSTAINABLE REFURBISHMENT OF EXISTING BUILDINGS FINANCIALLY VIABLE

Nigel Addy

Developers are looking to get the most out of their office stock and acquisitions in difficult economic conditions by considering refurbishment rather than redevelopment.

REFURBISHMENT, WHY WOULD WE DO THAT?

There is a compelling argument as to why a developer would decide to build a new building rather than refurbish an existing one. A new building would give the developer the ultimate flexibility to design a building to suit their requirements. These needs could be functional, architectural, viability or probably a combination of all. It gives the developer the ability to design a building to suit the targeted tenant's requirements. These can be very different from each other. For instance, a financial tenant will have very different requirements in terms of space planning than a technology-based tenant.

Each developer will have an agenda but the overriding one will be viability and the ability to let the building to a tenant or a number of tenants to meet their needs, both short and long term.

There are a number of positive reasons to refurbish instead of redevelop and all of these will be taken into account by a developer in their viability studies and development appraisal studies. At least eight reasons are possible.

Reduced capital expenditure

The cost of the refurbishment of a building is a lot less than for a new building. The difference will be determined by the extent of the refurbishment but in general terms the reuse of the existing foundations and structure is the minimum that is retained in even the most extensive refurbishments and will be a saving against a new development. Anything else that is retained is therefore added value as long as the economic life of what is retained is in line with the developer's expectation for the length of a lease.

Speedy planning process

The major refurbishment of a building which includes increasing area, changing the appearance of a building or changing its use will attract the attention of the planning authorities. A light refurbishment or medium refurbishment will not necessarily involve the planners, therefore that period in a development cycle can be discounted and the product can be returned to the market that much quicker. However, if, for example, the windows are being replaced then there is likely to be a requirement to satisfy statutory energy conservation standards (e.g. UK Building Regulations Part L) and this could lead to further expenditure in other elements of the building to bring it in line with current regulations.

Quick to market

One of the major advantages of a refurbishment over a new build development is that the building can be returned to the letting market quickly. This will again be determined by the extent of the refurbishment. A light refurbishment with minimal change to the cores and the fabric of the building can be turned around in a matter of months. A medium refurbishment can take anywhere between six to twelve months. There are also opportunities to phase the works so that an incumbent tenant could stay. This may lengthen the period on site and increase costs but the revenue stream still exists and will help to make the refurbishment more viable.

Greater tax relief

In many countries there are tax relief benefits related to refurbishment, for example, in the UK as a result of Capital Allowances. This is discussed later.

Removal of 'planning gain' type requirements

A minor or medium refurbishment will be unlikely to attract contributions to local services or highways improvements which might be required for a new construction.

Sustainable solution – reduced carbon footprint

The reuse of any part of an existing building will provide some benefits from a sustainability point of view. The structure of a building has embodied carbon and therefore not demolishing a building will provide a benefit.

There is a demand for buildings to be refurbished

This is on a number of levels. First, from a sustainable point of view people do not want to be seen to knock something down to rebuild something similar as it is seen as a waste. Developers will tend to do anything they can to keep an existing building for many of the reasons discussed in this chapter. Second, there is a demand in the marketplace for buildings that are different, quirky and reference their past. This is discussed later in this chapter.

Unlocking hidden value

There are many ways to unlock the hidden value in an existing building and these are discussed later.

WHAT MAKES A BUILDING SUITABLE FOR RETROFITTING?

There are a number of factors which determine the suitability of a building for refurbishment and the extent the refurbishment would take.

Building orientation and massing

Not all buildings are suitable for refurbishment as they may have floor plans that are too deep; they do not 'show' themselves and therefore reorientation to their surroundings is required which may be unviable in a refurbishment.

Slab-to-slab height

The slab-to-slab height, which is the distance between the top of the structural slab on one floor to the top of the structural slab on the next floor, is a critical dimension as it determines the maximum floor to ceiling height. The UK's British Council of Offices (BCO) states a minimum floor to ceiling height of 2.6 metres. As soon as this dimension falls below this mark it will affect the rental value of the floor. The BCO guide is something that is used as a standard for UK office buildings and agents compare a building on the market with these standards. If a building falls below then it is much more difficult to demand a high rent and to maximise the viability of a project. There are ways to maximise the floor to ceiling heights including removing ceilings and changing the servicing strategy, or reducing the raised floor (if there is one). With the use of less data cabling below floors there is a reduced need for a large void under floors and therefore a raised floor can be reduced from, say, 150 mm to 100 mm (clear 68 mm). There are further considerations such as how this affects thresholds to the core and the stairs but an increase in floor to ceiling heights would be of greater benefit than a few ramps to the cores.

Structural grids and floor loadings

Many buildings have been extended or refurbished in their lives and as a result the structures of some buildings can resemble a forest of columns. A new building has the advantage of starting with a blank sheet of paper and the structural grid is maximised through the design stages. An existing building may have a grid which does not suit a modern office requirement and it is prohibitively expensive (and risky) to remove the columns, therefore it is not viable and demolition is the answer.

Vertical circulation and services distributions

Getting services up and down a building is a primary concern when refurbishing a building and all the services are being renewed. The relocation of major plant may help or hinder the use of the existing risers. In general it is best to reuse existing risers as there will be no loss of floor area and therefore no loss of rental income for that floor area. Additional risers may be required for future tenant use, such as a kitchen extract. If the density of occupation of a space is to be increased from, say, 1:10 to 1:8, additional riser space may be required for the increase in the size of the ductwork providing fresh air. These considerations can add considerably to the cost of a refurbishment and hence its viability. The change of density will also affect the lift capacities, toilet and fire escape requirements and these need to be taken into consideration as a whole and not individually, as one decision affects another.

Relationship to neighbours

The desire to increase the size of an existing building and maximise the value needs to consider the effect this has on the neighbours and this could also affect the viability of a project. The 'Rights to Light' that the neighbours will be entitled to will limit the ability of a developer to develop a bigger building without paying what can be considerable sums to get rid of a Rights to Light issue. Neighbours can be very influential on the development, whether it is new or a refurbishment.

Floor plate width

Lower cost natural ventilation is a viable option for floor plates that are greater than 12 metres but not greater than 14 metres, and floor plates that are too deep inherently let for less money. There is a trade-off of value and floor area – a too dark space may let for little but by adding light via, for example, an atrium, may increase the value of the space compared to the cost of the atrium.

Unrealised value – hidden 'classic car' value

There are many examples of the refurbishment of commercial buildings which has uncovered the joy of an existing building. One of the best examples is the refurbishment of St Pancras station in the UK with its cast-iron columns and expansive glazed roof which have produced a cathedral-like feel to the space.

So many existing buildings have their own identity which can produce extraordinary design solutions which have added value to a property. Enjoying the heritage of a building is a very sustainable proposition and is a valuable asset of a building which can realise real value.

DIFFERENT LEVELS OF REFURBISHMENT MAY BE APPROPRIATE

The appropriate level of refurbishment will primarily be driven by the length of lease and the desired extension of a building's economic life, in addition to the current condition of the building. There are several key cost and value drivers to be considered.

Building condition/risks – structural constraints, asbestos, concrete repairs, etc.

The condition of the existing building will have a major impact on the viability of a scheme. The reuse of the structure has inherent risks and the mantra of survey, survey and survey again, can reduce risk and confirm the viability of a scheme. The best solution is a vacant building; however, this may not be possible. The minimisation of the extent of intervention into the structure of an existing building will reduce risk and uncertainty which all comes at a cost. The passing of the risk to those best able to manage it is the equitable way to deal with the unknown in an existing building.

Asbestos is a common problem in an existing building. A refurbishment/demolition survey will identify where asbestos is or is likely to be and the ideal solution is to clear the asbestos from a building before any work starts.

The condition of an existing building leads to a key decision on whether to refurbish or not. The economic life of the foundations and structure need to be taken into consideration whenever a refurbishment is to be undertaken. A key question from the tenants' due diligence team will be the life expectancy of the structure, how it has been dealt with and what are the inherent risks that will either need to be dealt with through the lease agreements or the rent applied.

External walls – repair or replace?

The choice to repair or replace the external walls of a building is another major decision in the refurbishment of an existing building. The intervention can be as little as repair and painting of windows to a full removal of the cladding and replacement. As the external walls are generally between 20 to 35 per cent of the cost of a new building the decision to replace the external walls is a key one. The decision will have an effect on what Building Regulations are required to be adhered to.

Mechanical and electrical services strategy

This is a key decision for a refurbishment programme and the key driver for the decision-making will be the economic life of the services. The figures below show the relative life of each major service in a commercial building.

An example of the decision-making process on changing services could be as follows:

A series of floors have been returned to the landlord at the end of a lease. The lease was for ten years and the building was new at the start of the lease. The office has a fan coil system to provide heating and cooling. The fan coils have been well maintained and are expected to last fifteen to twenty years as long as they are maintained properly. Therefore, they have a five- to ten-year life. Does the landlord renew or refurbish the units? If they are refurbished then the question of the repair and maintenance of the existing units needs to be taken into consideration either in the lease (e.g. excluded from the repair obligations) or the rent (maintaining full repair obligations). If the units are replaced then does the additional rent for the cost of a new system increase the return on the investment? There is a delicate balancing act of the cost of the options available to the developer and the viability of the refurbishment of the floors.

The reuse of the existing plant may keep inefficient plant running versus new plant which is much more efficient. This is a decision based on a cost-benefit exercise but will always be a sustainable decision as new plant is always more efficient than older plant.

Figures 4.1 and 4.2 show examples of the systems which are likely to need to be replaced based on extending a building services life by five or ten years. These types of tables help in the decision-making process when weighing up the balance of viability versus cost.

Mechanical Plant

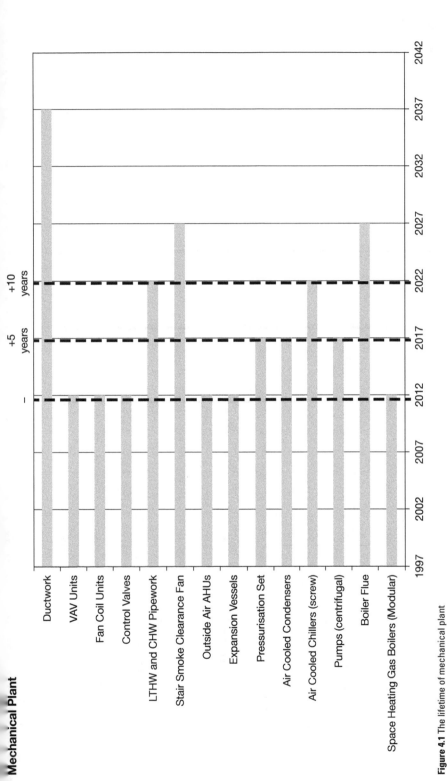

Figure 4.1 The lifetime of mechanical plant

Source: AECOM

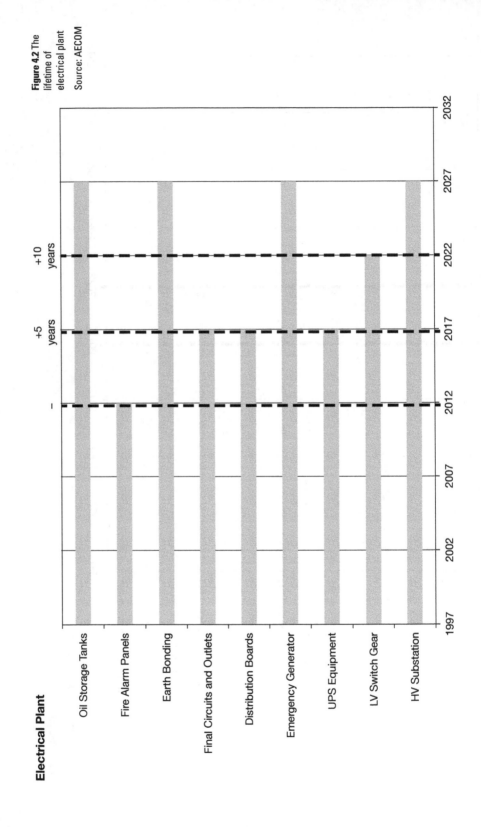

Electrical Plant

Figure 4.2 The lifetime of electrical plant

Source: AECOM

Layout and potential for extension/floor plate optimisation

Optimising the net lettable area of a building is the prime driver in a commercial building development. It drives the viability of a project. An existing building always has potential for some sort of gain in area whether through roof extensions, floor plate extensions, additional floors or infilling of atriums.

What occupiers want and what they can afford

The requirements, however, for different sectors of the market will also dictate the type of building that is required. Institutional tenants such as banks, lawyers and financial businesses are perceived to need a corporate feel to their buildings and this dictates both design and cost of the building. For example, in the UK, a refurbishment in the City of London will be very different to one in the West End or the fringes of these areas and the requirements of the various parts of London change constantly. The coming of the new East–West rail link, Crossrail, for instance, will define very different areas of London for development and various villages will spring up around the transport network.

There is a tipping point at which the refurbish costs are not viable and the refurbishment does not provide the return required for the development. The major factor determining this tipping point is the rental return for the space. This is determined by the quality of the space, location, etc. A refurbishment by its very nature will attract a different rent depending on the extent of the refurbishment. It is unlikely that the building will look new; however, a building can be transformed such as Case Study 4 (see Chapter 15), the Angel Building in London, whereby it looks like a new building and can attract rents which are very similar to new build space rents.

VALUE-BASED DECISIONS

The basic question is, does a change add value and is that commensurate to the additional rent that can be achieved? The refurbishment and reinvention of existing office stock offers a number of benefits when compared to complete redevelopment. Refurbishment enables a developer to achieve the following:

- The reuse of an existing asset.
- A better balance of risk and return.
- Quick delivery back to market (or refurbish whilst in use). Refurbishment offers a programme advantage relative to new build. Depending on the level of refurbishment, it is approximately 15–70 per cent quicker than new build.
- Maximisation of the value of an existing asset. With refurbishment, the developer is in the position where the unique style

and character of an older building can be retained. This can be useful when trying to market the space to businesses that are trying to differentiate rather than emulate. In addition to this, various helpful attributes of the original building, for example, car parking allocation and permitted development density and massing, can be retained.

- A more affordable approach, refurbishment can avoid the reconstruction of major structural elements and still retain the benefit of the creation and provision of new office space. Depending on the level of refurbishment, the cost of the refurbishment is approximately 10–75 per cent less than new build.
- Operational savings whilst re-energising the asset.
- Creation of an opportunity to support new ways of working.
- Potential reduction of the carbon footprint of an existing building. There is a strong argument relating to sustainability that supports refurbishment. The reuse of the majority of an existing building's fabric and an improvement to the building's services and performance may result in the reduction of the overall environmental impact when compared to a new build.

Not all of the existing office stock is suitable for refurbishment and there are certain factors that constrain and determine both the level and suitability of a building for refurbishment.

Creating and adding value through refurbishment

A carefully thought through refurbishment can transform a tired, uneconomical, inefficient building into a desirable, vibrant, efficient and profitable asset that supports new ways of working and incorporates modern sustainable technology.

The key to creating and adding value through refurbishment is to understand the factors that will attract tenants. These include flexibility, good floor to ceiling heights (greater than 2.5 metres), column-free spaces, the maximising of usable space, welcoming entrances, natural light and the right environment with adequate cooling.

The cost versus value equation is the primary driver for decision-making during the design stage of a refurbishment project. The creation of more valuable space as well as additional space is a primary consideration when determining the extent of any refurbishment.

Some examples of the sort of intervention to the existing building which can create more valuable space include:

- Removing columns, cores, staircases and other structural elements to open up usable space.
- Infilling or reducing existing atriums, courtyards, etc. to create additional usable space – an excellent example of this is the

Angel Building in Islington, London which resulted in a significant increase in usable floor area.

- Adding floors – this can be a relatively economic solution for creating additional area. The use of lightweight material can create light, open, column-free space which will be of increased value as it is new space usually located in a prime location on the top of the building. The amount of additional area can be constrained by the structural and service capacity of the existing building, rights of light issues and planning. The additional space also helps to create a 'blended product' that is ideally suited towards a multi-let where tenants that have a fixed requirement in terms of maximising floor to ceiling height and the height of the floor void can be accommodated on the new floors whilst those that are more flexible in this regard can be accommodated on the refurbished floors. This premium space should attract higher rental values.
- Reconfiguration of the core in a major refurbishment – the reconfiguration of the core or even the creation of a new core can provide additional area, greater efficiency, additional plant or rise space and better circulation.
- Removal of floor screeds to provide deeper floor voids – this may, however, create issues with fire separation, acoustics and air leakage.

The creation of a generous floor to ceiling height is a key consideration for a refurbishment. Relatively small floor voids are acceptable if the floor to ceiling heights are maximised. With the advance in technologies the need for larger floor voids is not as necessary; however, this does need to be considered against the tenant market that is being targeted.

The choice of the service strategy also needs to be carefully thought through. Many older buildings do not have the floor-to-floor heights to accommodate a four-pipe fan coil system and a suspended ceiling, whilst simultaneously maintaining an optimal floor to ceiling height.

The use of chilled beams, chilled ceilings and cooling mats are some of the ways to create the right environment whilst maintaining floor to ceiling heights. The use of displacement air systems in existing buildings is popular and enables users to open windows and provide natural ventilation during the summer months. This choice will be largely dependent on the ability to have a suitable floor void to deliver the air to the space.

The creation of a welcoming entrance should be a major consideration in any refurbishment of any scale. The first impression of the building is the entrance and a high-quality entrance will enhance the marketability of the building. Examples of how entrance areas can be altered to create a more inviting environment include:

- widening the entrance;
- creation of greater volume by forming double-height space;

- introduction of greater transparency and light;
- improved circulation and signage;
- improved disabled access, health and safety and fire regulations to meet current standards.

The refurbishment of common areas such as toilets and lift lobbies is also essential to a successful refurbishment. The extent of the refurbishment can range from a refresh of finishes, which can be done relatively cheaply, to a full reconfiguration of the toilet areas including replacing sanitaryware, finishes and services.

The complete refurbishment of a building provides an opportunity to review the services strategy. The ability to move major plant from, say, the roof to basement can release area which can be turned into valuable usable space. The freeing up of area on a roof can enable the potential to build additional floors.

Finally, refurbishment can also add value by allowing the opportunity to provide additional environmental facilities to achieve additional Building Research Establishment Environmental Assessment Method (BREEAM) points. This is an important consideration with a major refurbishment. A BREEAM rating of excellent can be achieved in a refurbished building and this increases the marketability of the existing asset.

Sustainability

It is generally considered that the retention of an existing building is considered more sustainable as it is using existing building stock on 100 per cent previously developed land.

As part of a new building design process, measures to incorporate sustainable features go hand in hand. With refurbishments the process is a little more challenging. Sustainable features for consideration in refurbishment schemes include:

- Low water use equipment.
- Introduction of lighting controls and the extensive metering to monitor the performance of the building and occupants.
- Green walls and roofs to attenuate rainwater and aids in the sustainable drainage design.
- External shading and facade treatment can assist in the reduction of energy consumption; however, methods available such as introduction of new glazing and external shading are restrictive in refurbishments. The introduction of secondary glazing can be considered for a refurbishment.
- Refurbishment projects can lend themselves to innovative services solutions which ordinarily would not be acceptable in the modern new corporate building. Exposed services and use of thermal mass for promotion of cooling, for example.

- Natural ventilation and mixed mode environmental control. This will be influenced by the size, location and shape of the floor plate.

Due to the nature of refurbishment projects the sustainability solutions will be different for each building and dependent on the constraints of the existing building. The cost of providing a sustainable solution needs to be considered as a whole-life cost as well as a cost versus value decision. A cost-benefit analysis should be carried out to give a monetary value for making a decision; however, other factors will also influence which sustainable design decisions are made. For example, the additional capital cost for providing the requirements of BREEAM or Leadership in Energy and Environmental Design (LEED) can be offset as the marketability of the finished building will be greater than a lower sustainability rated building and possibly a greater rental value.

The whole-life cost of a solution will need to be considered, i.e. the capital cost and the running and maintenance costs over the life of the product to establish the benefits of any given solution. For example, the use of natural ventilation will save a great deal of money compared to a full air-conditioned building, both in capital and running costs. It also has the added advantage of being more appealing to the user.

Another example of creating value from sustainable solutions would be the use of green roofs, coupled with introducing balconies and access to the roofs. This is not only a low cost but adds amenities to the tenants which can be included in an increased rent.

COST MODELLING

For buildings offering limited potential for sustained uplift in value, a small refurbishment might provide an opportunity to enhance rental income ahead of a comprehensive redevelopment. A more extensive refurbishment will close off the opportunity for redevelopment for an extended period. On the basis of this, a developer does need to consider the available opportunities. The options open to a developer can be defined as in Figure 4.3.

The indicative costs of refurbishment given in Figure 4.3 are driven by the design and condition of the building, constraints on the design and construction solutions adopted as well as by the scope of works required to meet the project objectives. All projects are unique, and as a result, cost ranges are much broader than for new build. Furthermore, substantial allowances for additional costs associated with design development and risk should ideally be retained in project budgets until a late stage in the programme, to take into account the potential for further changes in scope based on better knowledge of the existing building. (The figures exclude any occupier enhancements and furniture installation.)

Table 4.1 defines the extent of each type of refurbishment.

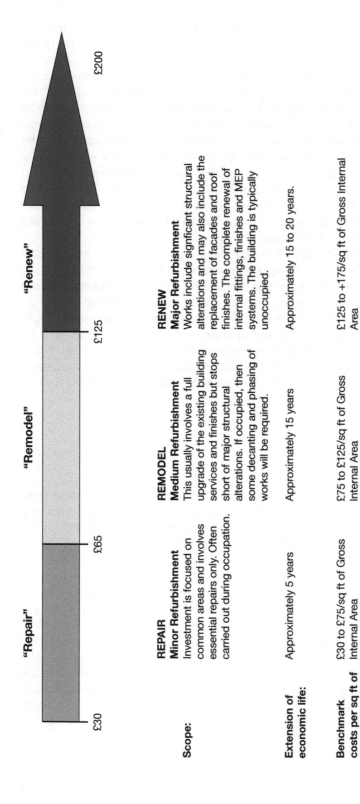

"Repair"   "Remodel"   "Renew"

£30   £65   £125   £200

	REPAIR Minor Refurbishment	REMODEL Medium Refurbishment	RENEW Major Refurbishment
Scope:	Investment is focused on common areas and involves essential repairs only. Often carried out during occupation.	This usually involves a full upgrade of the existing building services and finishes but stops short of major structural alterations. If occupied, then some decanting and phasing of works will be required.	Works include signficant structural alterations and may also include the replacement of facades and roof finishes. The complete renewal of internal fittings, finishes and MEP systems. The building is typically unoccupied.
Extension of economic life:	Approximately 5 years	Approximately 15 years	Approximately 15 to 20 years.
Benchmark costs per sq ft of Gross Internal Area:	£30 to £75/sq ft of Gross Internal Area	£75 to £125/sq ft of Gross Internal Area	£125 to +175/sq ft of Gross Internal Area

Figure 4.3 Levels of refurbishment with example 2013 costs for central London location

Source: AECOM

	Repair Minor Refurbishment	Remodel Medium Refurbishment	Renew Major Refurbishment
Substructure	Isolated repairs to basement area waterproofing	Repairs to basement area waterproofing and local repairs to structure	Replace and reconfigure basement space and foundations
Frame, Upper Floors and Stairs	Isolated repairs to structure, clean and paint walls, soffits and balustrades to stairs	Minor remodelling of structure, clean and paint walls, soffits and balustrades to stairs	Significant replanning to walls and floors, incl infilling atria, new finishes and balustrade to stairs
Roof	Isolated repairs where necessary	Isolated repairs with limited replacement of finishes	Replace roof coverings, new atrium roof if necessary
External Walls and Windows	Clean only and isolated repairs	Minor repairs, replace doors, reception remodelling	Repairs, replace doors, infilling/renewing sections of the facade
Internal Walls and Doors	Make good general wear and tear, minor reconfig and repair doors	Remodel reception, reconfigure core and Reception. Replace front of house doors	Significant replanning of core and Reception. Replace all doors
Finishes and Fittings	Patch repairs/works to enhance finishes. Paint finishes	Replace existing finishes to landlord areas, incl Reception.	Replacement of all finishes and fittings incl new sanitaryware
MEP	Complete overhaul of all existing plant to allow 10 yrs life.	Replacement of all existing plant and distribution systems	Complete replacement of all new services to allow 15–20 years life
Lift Installation	Overhaul existing lifts and refurb lift cars	New lift cars in existing shafts with new motors and controls	Complete replacement of lifts and lift motor room
Cat A Fit Out	Strip out existing cellurisation, repairs to floors and ceilings and minor reconfig of services	Strip out existing cellurisation, replace floors and ceilings and modify/upgrade services	Strip out existing cellurisation, replace floors, ceilings and services

Source: AECOM

REFURBISHMENTS – TAX ALLOWANCES

With the economic climate within the building industry in 2014, and the more recent period of recession in Europe, it is understandable that there is likely to be more refurbishment of neglected commercial property undertaken in the near future. Legislative pressures around reduction in energy use in buildings, together with rising energy costs, put pressure on landlords and occupiers alike to improve their building stock. These sticks often act as an impetus, but it is also useful to be aware of the carrots in the form of tax incentives in different countries that can be very significant within refurbishment works.

Most countries will provide tax allowances to account for depreciation of assets. These include capital allowances, wear and tear allowances, cost segregation and depreciation regimes all with minor differences due to local rules and the use of these incentives by the applicable countries' governments to suit their particular ends. They will generally give the investor the ability to depreciate the cost of certain assets within the property over a number of years on a straight line, or reducing balance basis. In some instances there will also be a depreciation allowance on the building structure itself. In most cases the higher depreciation rates (or the only depreciation rates) will be for items of interior fit-out or plant and machinery.

Therefore, with refurbishment works the majority of the costs will tend to fall on items attracting higher levels of depreciation or allowance. In some instances tax relief will be applicable in some form to 80 per cent or more of the costs of a refurbishment, meaning the tax savings in the future will help to offset the capital costs. For instance, if 80 per cent of the cost attracts tax relief and there is a 30 per cent tax payable by the investor it will be possible, over time, to reclaim 24 per cent of the costs in tax relief.

Countries are also using their tax regimes to encourage investment in energy-saving and other environmentally beneficial technologies. With Building Regulations and running costs requiring more energy-efficient installations and renewable energy being targeted for many new projects there are often higher allowances for these types of investments. A typical example would be the UK's Enhanced Capital Allowances regime for energy- and water saving technology installed within buildings. It offers a 100 per cent first-year allowance against the costs of these installations. Therefore, it improves the cash flow of the tax relief so the investor is able to obtain their tax savings at an earlier date.

The benefit of these allowances needs to be considered as part of the appraisal for the whole project so that the feasibility is based on a holistic evaluation inclusive of all costs, income and benefits including the future tax relief. It is often the case with these allowances that the values are not limited to the qualifying refurbishment costs only. They can include associated installation costs, contractor's establishment and on costs as well as relevant professional fees. The calculations will often require input from a specialist surveyor who will be trained in building technology, construction costs, relevant tax legislation and case law.

CONCLUSION

In conclusion, the refurbishment of an existing building rather than demolishing and rebuilding can have many benefits. The major benefits are a reduced cost of construction, a return to the market as quickly as possible at an enhanced value, the reuse of an existing asset which in itself is sustainable and the rediscovering/reinventing of a building through refurbishment.

5
THE CURRENT ENERGY PERFORMANCE OF COMMERCIAL BUILDINGS IN NORTHERN CLIMATES: EUROPE

Dan Staniaszek

In this chapter, the energy consumption characteristics of the commercial building sector in Europe are described. These are put into the context of both the regulatory environment, primarily European Union (EU) legislation, and the commercial drivers for renovating existing buildings. Aggregate results from a major multi-annual refurbishment programme are presented, before the requirements for new construction are summarised.

EUROPE'S COMMERCIAL BUILDING SECTOR

The commercial sector in Europe is characterised by a great diversity in types, styles, sizes, ages and uses, as well as the climatic and cultural context within which buildings are constructed. The desire to retain the rich European cultural heritage whilst developing thriving, post-industrial economies means that towns and cities across Europe display all manner of building types, old and new, in myriad combinations. Many seemingly old buildings in city centres are in fact new constructions, where only the facade has been retained.

In order to generate a clear picture of the state of play of Europe's built environment, and in particular its energy performance, the Buildings Performance Institute Europe (BPIE) undertook perhaps the most extensive survey of its kind in 2011 across twenty-nine countries – the twenty-seven Member States of the EU at the time ('EU-27'), together with Norway and Switzerland.[1] Key findings include:

- The total built up floor area across the twenty-nine countries in the survey amounts to 25 billion m^2.
- Three quarters of floor area is accounted for by residential properties; commercial and public buildings (i.e. non-residential buildings) make up one quarter of the total, or just over 6 billion m^2.
- Energy use in buildings accounts for 40 per cent of all energy use in Europe.
- Energy use per unit floor area in non-residential buildings, at 280 kWh/m^2/year, is more than 40 per cent greater than in residential buildings. This is due to higher service energy use, for example ICT equipment in offices, specialist medical equipment in hospitals and typically greater use of air conditioning and lighting across most building types.
- Whilst non-residential buildings account for only 25 per cent of the total built area, their higher energy use per unit floor area mean they account for around one third of total energy use.
- Within the non-residential sector, 12 per cent of the stock is in public ownership. The split between public and private (i.e. commercial) ownership is not generally available on a country-by-country basis.

Europe can be divided into three broad regions, based on climate, economic development and/or building characteristics:

- Northern and Western Europe, which is the largest, most populous and wealthiest region, extending from the Atlantic coast in the west, the Alps in the south, Germany and Austria in the east and northwards to encompass Scandinavia. Half of the total built area is found in this region, inhabited by a population of 281 million, or 55 per cent of the total. All countries in this region have Gross Domestic Product (GDP) per capita (in purchasing power parity, or PPP, terms) above the EU average.
- Southern Europe includes mainly countries on the northern fringes of the Mediterranean Sea (Spain, Italy, Malta, Cyprus, Greece), together with Portugal on the Atlantic coast. These

countries together comprise over a third (36 per cent) of the built area, but only 25 per cent of the population. The GDP PPP for these countries ranges from 75 per cent to 99 per cent of the EU average.

- Central and Eastern Europe is the smallest of the three regions, both in terms of built area (14 per cent) and population (20 per cent). The ten member states in this group are all former communist countries, and as such display a degree of similarity in the nature of the building stock construction, notably in the period to 1990. For the most part, these countries are less developed economically, having a GDP PPP between 50 per cent and 80 per cent of the EU average.

To demonstrate the climatic range within the EU, Figure 5.1 shows the commonly used measure of heating degree days (HDD) to illustrate the severity of the weather, by summing, over a year, the difference between the average daily external temperature and a nominal internal temperature. The higher the number, the greater the energy required to maintain comfortable temperatures, all other things being equal.

The climatic range across Europe is significant, with countries in the south region having HDD less than 2,000 while Scandinavian countries exceed 5,000 HDD. This has a clear bearing on energy usage for heating (the most significant end use requirement across Europe, accounting for 45 per cent of the total; see Figure 5.2), varying as it does from under 10 per cent of total energy use in Malta to 60 per cent in Slovakia.

While important, external temperature is not the only factor affecting energy use. Many office buildings in Europe today need little additional heating, due to the internally generated heat from lighting, appliances and occupants themselves. The need to avoid overheating in summer is often more of a challenge. That said, energy used for cooling in the commercial sector represents just 7 per cent of the total as a European average, but reaches 25–30 per cent in the warmest countries – Cyprus and Malta.

Understanding energy use in the non-residential sector is complex, since end uses such as lighting, ventilation, heating, cooling, refrigeration, IT equipment and appliances vary greatly from one building category to another. One of the most striking features about the commercial sector is that, over the last twenty years, electricity consumption has increased by a remarkable 74 per cent, as depicted in Figure 5.3 (dashed line). This is the result of an increasing penetration of IT equipment, air-conditioning systems, etc. Note, however, that overall use of fuels other than electricity (solid line) has remained more or less constant over the period, despite increasing total floor area.

In terms of use, trade premises and offices are the dominant sectors, together accounting for over half the non-residential building floor area. Other important sectors include education, hotels and restaurants, hospitals and sports facilities, as demonstrated in Table 5.1. The building type breakdown by country is presented in Figure 5.4.

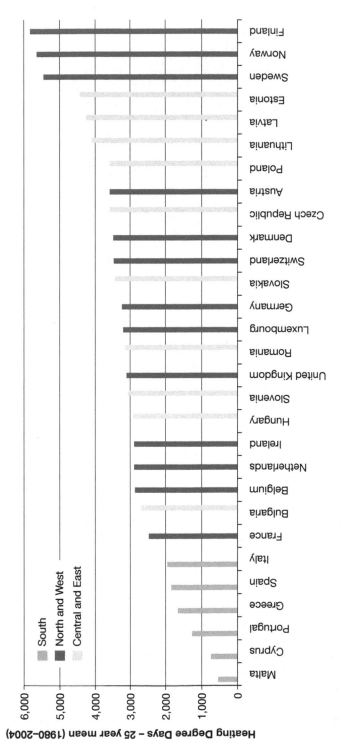

Figure 5.1 Degree day spread across EU-27, Norway and Switzerland

Source: Eurostat

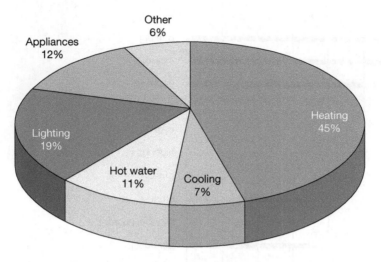

Figure 5.2 End use energy requirements in European non-residential buildings

Source: BPIE

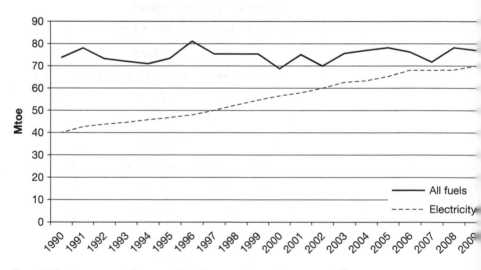

Figure 5.3 Final energy use in the non-residential sector – EU-27 + Norway and Switzerland

Source: BPIE and Eurostat

The variation in specific energy use across different commercial building types is illustrated in Figure 5.5. Within the countries for which data are available, hotels and restaurants and hospitals generally have the highest energy use per unit of floor area, with educational buildings and sports facilities the lowest. That said, there is considerable variation between the different countries.

Table 5.2 puts non-residential energy use into context. Compared with the other main sectors (transport, industry and residential) the non-

Table 5.1 Building stock by non-residential building sector (EU-27, Norway and Switzerland)

Sector	Aggregate floor area (million m^2)	Percentage of total
Wholesale and retail trade	1,670	28%
Offices	1,386	23%
Educational buildings	1,018	17%
Hotels and restaurants	664	11%
Hospitals	396	7%
Sports facilities	264	4%
Other	660	11%
Total	6,056	100%

residential, or services, sector is the smallest, accounting for 13 per cent of EU-27 final energy use in 2010. However, it is by far the fastest growing, having increased by 32 per cent in a decade, against an overall increase, in all sectors, of just 3 per cent. What is more striking is that all of the increase in total energy use is accounted for by the service sector, since aggregate energy use in the remaining sectors decreased slightly in the period 2000–10. For this reason, it can be seen that increasing the efficiency with which energy is used in the services sector is key to meeting the EU's environmental goals.

EU POLICY CONTEXT

As a significant contributor to EU energy consumption, resource utilisation and carbon emissions, the building sector is subject to numerous policies, strategies and long-term goals which seek to reduce its impact. These have been formulated into the so-called '20-20-20' targets, which is a set of three key objectives for 2020:

- a 20 per cent reduction in EU greenhouse gas emissions from 1990 levels;
- raising the share of EU energy consumption produced from renewable resources to 20 per cent;
- a 20 per cent improvement in the EU's energy efficiency.

Looking out across a more distant horizon, the EU has a set of longer term objectives, set out in roadmaps to 2050. As far as the building sector is concerned, the three principal ones are:

- *EU Roadmap for Moving to a Competitive Low-Carbon Economy in 2050* (European Commission, 2011), which identified the need of reducing carbon emissions in residential and services

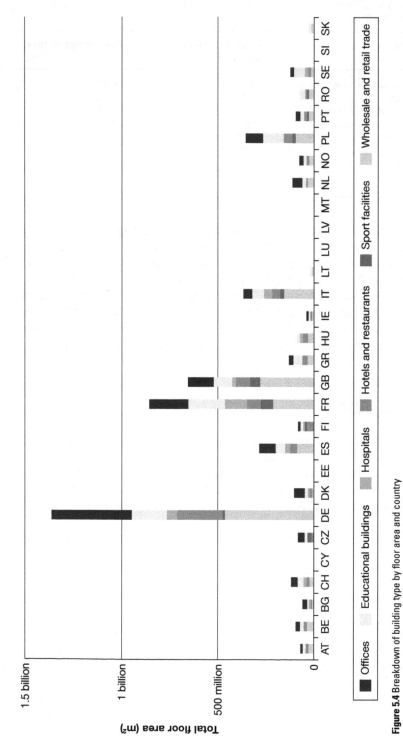

Figure 5.4 Breakdown of building type by floor area and country

Source: BPIE database (www.buildingsdata.eu, accessed 2 July 2013)

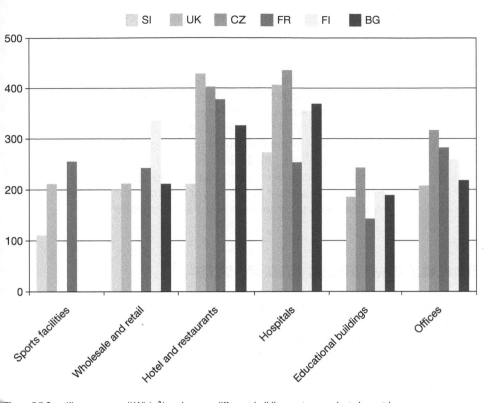

Figure 5.5 Specific energy use (kWh/m²/year) across different building sectors – selected countries
Source: BPIE

Table 5.2 EU Final Energy Consumption by Sector (million tonnes of oil equivalent, Mtoe)

	2000	2001	2002	2003	2004	2005	2006	2007	2008	2009	2010	Change (%)
Total	1,121	1,145	1,132	1,171	1,186	1,191	1,192	1,165	1,174	1,112	1,153	3
Industry	330	329	325	338	336	331	324	323	313	268	292	−12
Transport	341	345	348	353	363	367	374	380	378	367	365	7
Residential	292	302	293	298	302	303	300	285	297	294	307	5
Services	115	127	125	131	134	136	139	136	144	143	152	32
Other	42	43	41	51	51	54	54	42	43	40	37	−13

Source: 'Energy, transport and environment indicators', Eurostat, 2012

sectors (collectively, the building sector) by 88–91 per cent by 2050 compared to 1990 levels.

- *Energy Roadmap 2050* (European Commission, 2011) states that 'higher energy efficiency potential in new and existing buildings is key' in reaching a sustainable energy future and contributing significantly to reduced energy demand, increased security of energy supply and increased competitiveness.
- *Roadmap for a Resource Efficient Europe* (European Commission, 2011) identified buildings among the top three sectors responsible for 70–80 per cent of all environmental impacts. Better construction and use of buildings in the EU would influence more than 50 per cent of all extracted materials and could save up to 30 per cent of water consumption.

EUROPEAN DIRECTIVES AFFECTING THE BUILDING SECTOR

At the other end of the spectrum are the various regulations and directives that apply across all Member States as current requirements affecting the building stock. These provide a common framework within which individual Member States are required to set standards and performance levels regarding energy use in buildings. For the most part, these regulations apply equally to commercial, public sector and residential buildings alike. The main directives are:

- the Energy Performance of Buildings Directive (EPBD), originally introduced in 2002, and recast in 2010 (2010/31/EU);
- the Energy Efficiency Directive (EED) 2012/27/EU, introduced in 2012;
- the Renewable Energy Directive (RED), 2009/28/EC, introduced in 2009.

RED requires the use of minimum levels of energy from renewable sources in new buildings and in existing buildings that are subject to major renovation. The main provisions in EPBD and EED, as they relate to sustainable retrofitting of buildings, are described in more detail below.

Energy Performance of Buildings Directive (EPBD) – main provisions

The first major attempt to set a European framework for the energy performance of buildings came in 2002 when the EPBD set out a number of requirements on Member States, ranging from the establishment of certification schemes for buildings (so-called energy performance certificates, or EPCs), inspection regimes for major heating and air-conditioning plant, and performance requirements on Building Regulations. In many ways,

EPBD simply raised the bar in terms of standards across all EU Member States to the performance of some of the best. For example, Denmark and the Netherlands had already established certification schemes for buildings, and EPBD required other Member States to introduce similar mechanisms.

Whilst the original EPBD made good progress in a number of areas, implementation at Member State level was slow and incomplete, while some of the provisions were not having the desired effect in terms of achieving energy savings. With that in mind, the European Commission commenced a review in 2009, resulting in the recast Directive introduced in 2010. EPBD now contains the following main provisions:

- Methodology for calculating the energy performance of buildings and setting of minimum energy performance requirements – Member States are required to apply a common methodology for calculating the energy performance of buildings, and set minimum energy performance requirements at cost-optimal levels, using a comparative methodology framework developed by the Commission.
- Requirements for nearly zero-energy buildings (nZEB) – from the end of 2020, all newly constructed buildings will have to consume 'nearly zero' energy, with the low level of energy coming 'to a very large extent' from renewable sources. For buildings occupied and owned by public authorities, this requirement must be met two years earlier. Furthermore, Member States are required to prepare national plans for increasing the number of nZEB, across new and existing building stocks. These plans may include targets differentiated according to building category.
- Requirement to improve the energy performance of existing buildings undergoing major renovation – the recast directive extended the scope of the original EPBD to almost all existing and new buildings and removed the previous 1,000 m² threshold for major renovations that existed in the 2002 EPBD (this threshold had excluded 72 per cent of the building stock). When existing buildings undergo 'major renovation', their energy performance should be upgraded in order to meet minimum energy performance requirements.
- Technical building systems – in order to optimise the energy use of technical building systems such as heating, ventilation and air-conditioning (HVAC) plant and lighting systems, Member States need to set system requirements in respect of the overall energy performance, the proper installation, sizing, adjustment and control of such systems which are installed in existing buildings. Member States may also apply these system requirements to new buildings.
- Financial incentives and market barriers – Member States are required to review and publish details of existing and proposed

measures/instruments, including those of a financial nature, which address market barriers and which seek to improve the energy performance of buildings and aid the transition to nZEBs.

- Energy Performance Certificates – EPCs are required to be issued for all buildings when sold, rented, or newly constructed. For certain larger buildings visited frequently by the public, these certificates must be displayed in a prominent place. Annual reports on the quality of EPCs need to be produced by the relevant authorities with responsibility for implementing the control system.
- Inspection of heating and air-conditioning systems – larger heating and air conditioning systems need to be inspected on a regular basis. These inspections must be undertaken by suitably qualified experts, and a report issued to the owner or tenant of the buildings after each inspection. This report must include recommendations for the cost-effective improvement of the energy performance of the inspected system.

Energy Efficiency Directive (EED) – main provisions

Whilst the EED takes a wider perspective across all end uses and not just the building sector, a number of provisions are geared specifically towards encouraging the sustainable retrofit of buildings. These include:

- Building renovation – all Member States are required to set out strategies for the renovation of national building stocks, including commercial, public and residential buildings.
- Exemplary role of public bodies – national governments are required to show leadership in improving the energy performance of their building stocks by renovating 3 per cent by floor area of buildings owned and occupied by central governments every year. Furthermore, central governments are required to only purchase buildings (as well as products and services) with high energy efficiency performance.
- Metering and billing – measures to increase transparency and accuracy of energy costs are intended to raise awareness amongst building owners and occupiers as to the opportunities for saving money through improving the energy performance of buildings they own and/or occupy.

COMMERCIAL DRIVERS FOR SUSTAINABLE RENOVATION

In 2012, BPIE, in collaboration with the Global Buildings Performance Network, commissioned the Economist Intelligence Unit (EIU) to undertake a survey of building sector executives to determine their views on drivers for sustainable renovation of commercial real estate.[2] Key findings include:

- The financial crisis, which caused downward pressure on real estate valuations, highlighted the need for renovation of existing building stock in order to maintain and increase the value of portfolios. Furthermore, deep retrofits were found to provide a long-term benefit in achieving lasting value improvement.
- Compared with other global regions, EU companies are relatively active in retrofitting buildings, but efforts need to double to meet EU energy efficiency goals by 2020. Of EU respondents in the building sector 43 per cent focus on retrofits, compared with 37 per cent in the USA and 23 per cent in China. Despite this, energy-efficient retrofits still account for only 1 per cent of existing stock.
- The EU has taken some positive steps to improve regulation, but ambiguity regarding definitions of what constitutes a 'deep retrofit' and a 'nearly zero-energy building' affects implementation at national levels. Of respondents, 29 per cent identified regulatory uncertainty as a barrier to pursuing energy-saving investments. Furthermore, implementation of EU directives varies by country, which limits the ability of property owners to achieve economies of scale across the region.
- That said, regulatory uncertainty should not be used as an excuse for inaction, as this exposes companies to the risk of asset depreciation. Large property owners are starting to audit their portfolios to identify where they can achieve the most cost-effective measures. The deeper the retrofit, the lower the asset depreciation risk.
- Attracting large institutional investors in retrofit finance will require project aggregators. To be effective, aggregators require clear energy performance objectives, standard contract structures that allocate responsibility for performance, and data collection and transparency about results.

KEY RETROFIT OPPORTUNITIES

Every aspect of operation in commercial buildings provides an opportunity to cut energy use and make valuable cost savings. An ideal scenario is when a building is empty, for example, between tenancies, and alongside measures to improve the amenities or cosmetic look of the building. Rather than installing retrofit measures piecemeal, a holistic approach will provide the greatest opportunity to maximise savings, as this allows for integrated solutions. For example, design measures to cut summer heat gain, such as natural ventilation and solar shading, may result in eliminating the need for mechanical cooling, providing significant cost savings in HVAC plant (see Chapter 8).

The range of retrofit activities varies enormously, in terms of measures installed, savings achieved, cost and a host of other parameters. To give a flavour of opportunities, the results of the EU GreenBuilding Programme are summarised in this section.

GreenBuilding Programme

The GreenBuilding Programme is an initiative of the European Commission, launched in 2005, aimed at improving the energy efficiency and introducing renewable energy sources into non-residential buildings. Over the period to 2010, nearly 300 buildings were included within the programme from seventeen countries, of which more than half (163) were building renovations.

Some 60 per cent of buildings in the programme were offices, though the total included over a dozen building types, ranging from educational establishments and hotels to retail and leisure facilities. Building size ranged from 400 m² to 15,600 m², while the age profile (i.e. year of construction) ranged from pre-1900 to the present day.

On average, 3.5 measures were installed per building. Figure 5.6 shows the spread of measures, with HVAC, building fabric and lighting being the most common.

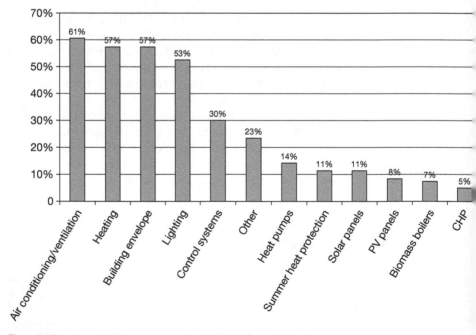

Figure 5.6 Prevalence of different measures installed in the GreenBuilding Programme

Source: EU Joint Research Centre

The primary energy demand in existing office buildings in the sample was reduced by an average of 41 per cent as a result of the retrofit measures. Individual projects saved between 26 per cent and 89 per cent. All projects were required to be cost-effective, achieving a positive Net Present Value. For those participants that reported Internal Rates of Return, these were in the range 9–20 per cent.

Details of the GreenBuilding Programme, which is administered by the Commission's Joint Research Centre (JRC), can be found at http://re.jrc.ec.europa.eu/energyefficiency/greenbuilding/ (accessed 2 July 2013).

REQUIREMENTS FOR NEW BUILDINGS

Given the wide variation in climate and historical approach to the setting of energy performance requirements for buildings across the EU, there is no standard method whereby individual Member States set energy requirements for new non-residential buildings. Historically, most countries adopted an elemental approach, setting maximum heat loss ('U-value') for individual building elements such as walls and roofs, though in recent years, the trend has been towards overall performance-based requirements.

Whilst the performance metric of energy use per unit of floor area ($kWh/m^2/year$) is the most widely used, it is by no means the only one. At least one country (Austria) uses heated/cooled volume, rather than area, while another (UK) uses a metric based on CO_2 emissions (kg $CO_2/m^2/year$). Some countries use primary energy as the basis, while others use delivered energy – variants range from heating energy use only, to all end uses. Some countries set two benchmarks, for example one for total final energy, and one for heating and cooling energy.

Table 5.3 summarises the prevailing requirements for new building energy performance across the EU-27, Norway and Switzerland. Not only is there a variety of ways in which requirements are expressed, but also there are significant differences between individual countries when the same metrics are compared. Therefore, it is difficult to draw clear conclusions regarding energy use in new buildings. That said, based on some of the more ambitious national requirements, a level of energy consumption for commercial and public buildings (other than hospitals, which have greater operational demand for energy) of 100 $kWh/m^2/year$ or less should readily be achievable today. This compares with the current average consumption level of 240 $kWh/m^2/year$.

Denmark, for example, has a current requirement for commercial and public buildings to consume no more than 73 $kWh/m^2/year$ (for a typical 1,000 m^2 building). Furthermore, the requirement to reduce this to 42 $kWh/m^2/year$ in 2015 and 25 $kWh/m^2/year$ in 2020 has already been announced, giving the industry time to prepare for these more demanding levels of performance.

Table 5.3 Prevailing requirements for new building energy performance across the EU-27, Norway and Switzerland

Country	Basis	Performance-based requirements for new buildings. NB All figures are kWh/m²/year unless otherwise stated						NOTES
		Offices	Educational Buildings	Hospitals	Hotels & Restaurants	Sports facilities	Wholesale & retail trade	
Austria	Heating & Cooling	VOLUME BASED REQUIREMENTS: 22.75 kWh/m³/year (heating); 1 kWh/m³/year (cooling)						RENOVATIONS: 87.5 kWh/m³/year (heating); 2 kWh/m³/year (cooling)
Belgium - Brussels		E75					E75 (services)	The E number in Belgium is a measure of building energy performance. Each region has its own set of regulations
Belgium - Wallonie		E < 100						
Belgium - Flanders		From 2012, E70 From 2014, E60						
Bulgaria	Final Energy	80–132	56–98	180–242	176–230	90–134	90–134	
	Heating & Cooling	40–82	40–82	50–102.5	50–102.5	40–82	40–82	
Cyprus		A or B category on the EPC scale						
Czech Rep.	Final Energy			179 | 130 | 310 | 294 | 145 | 183				
Denmark	Primary Energy	71.3 kWh/m²/year + 1650/A						'A' denotes gross heated floor area
Estonia	Primary Energy	220	300	400	300	300	300	
Finland		Based on thermal transmittance (heat loss) measured in W/K. Max. U-values for new buildings: Roof – 0.09; Wall – 0.17; Floor – 0.09; Window – 1						
France	Primary Energy			50				
Germany	Primary Energy	New buildings must not exceed a defined primary energy demand for heating, hot water, ventilation, cooling and lighting installations based on a reference building of the same geometry, net floor space, alignment and utilisation.						
Greece	Primary Energy	The Primary energy requirement for new and renovated building in Greece is = 0.33 – 2.73 × Reference Building energy performance						
Hungary	Primary Energy	132–260	90–254					
Ireland		MPEPC (Max. Permitted Energy Performance Coefficient) should not exceed 1, equivalent to the boundary between B and C in the EPC scale.						
Italy	Heating, DHW, cooling and lighting	Buildings must achieve Class A + to C						

Country	Metric							Notes
Latvia	Final Energy							Min Class C buildings: 80 kWh/m²/year for buildings over 3000 m², 100 kWh/m²/year for buildings 501–3000 m², 115 kWh/m²/year for buildings up to 500 m²
Lithuania								$HTR = h_A * A$ (1–2 storeys: $h = 1.1$ W/m².k; 3–4 storeys: $h = 0.9$ W/m².k; $>= 5$ storeys: $h = 0.7$ W/m².k). Heat loss coefficient must be lower than HTR
Luxembourg								Requirements are based on building elements, rather than overall performance
Malta								Requirements are based on building elements, rather than overall performance
Netherlands		EPC = 0.6						EPC = Energy Performance Coefficient. In 2000, EPC = 1. The target is EPC = 0 (i.e. zero-energy building) in 2020
Norway	Net energy demand inc. lights/appliances	150	120–160	300–335	220	170	210	
Poland	Final Energy	174						Requirements for other non-residential buildings apply
	Heating & Cooling	183						
Portugal	Primary Energy	407	174	465	523/1395	233	1279	
	Final Energy	122	52	140	157/419	70	384	
Romania								Requirements are based on building elements, rather than overall performance
Slovenia	Heating & Cooling			30–90				
Slovakia	Total Delivered Energy		240	84	201	95	161	
					187			
Spain								The energy performance requirements is not expressed in units of kWh/m²/year
Sweden	Final Energy	(electrically heated buildings): 55–95 kWh/m²/year; (non-electrically heated buildings): 100–140 kWh/m²/year						Range depends on climatic zone
Switzerland	Heating	46	43	44	58	40	36	RENOVATIONS: 125% of new build requirements
United Kingdom		Target carbon dioxide Emission Rate (TER) values apply for non-domestic buildings. The UK requirements are based on achieving a % reduction in CO_2 emissions over a notional building of the same size/shape. The equivalent value is approx. 60 kWh/m²/year						

Source: BPIE survey

DAN STANIASZEK

THE PATH TO A SUSTAINABLE COMMERCIAL BUILDING STOCK

With most of today's existing buildings having been constructed in an era when energy use was not a significant concern, and given the technological innovation in products, construction techniques and solutions to capture energy from renewable sources in recent years, the potential for reducing energy use in buildings is significant. A number of countries have already set out their vision of a low-carbon building stock. For example, Germany aims for a reduction in primary energy use in buildings of 80 per cent by 2050, while France is aiming for all new buildings to generate more energy than they consume (so-called 'positive energy buildings') from 2020.

Achieving a sustainable building stock requires action on two fronts: renovation of existing buildings to high energy performance standards, and setting of progressively tougher performance requirements for new buildings – ultimately down to zero net energy use and positive energy buildings. Whilst EU directives set the scene for such a transition, it can only be achieved if stakeholders throughout the supply chain gear up to deliver high-quality renovations and new buildings with very low energy requirements, and if there is a demand for such buildings from owners and occupants.

NOTES

1   'Europe's buildings under the microscope', BPIE, 2011. Economic and climate data sourced from Eurostat.
2   www.gbpn.org/sites/default/files/06.EIU_EUROPE_CaseStudy_0.pdf (accessed 2 July 2013).

REFERENCE

European Commission (2011) *European Commission's Roadmap for Moving to a Competitive Low Carbon Economy in 2050*, Brussels: European Commission.

6
EFFECTING DESIGN PROCESSES AND PRACTICES FOR SUSTAINABLE REFURBISHMENT

Lizi Cushen

WHY STANDARD DESIGN PROCESSES NEED CHANGING TO PROMOTE SUSTAINABILITY

One of the main challenges currently facing the construction industry is the development of sustainable practices in order to reduce environmental impacts and improve social and economic aspects of refurbishment projects. A major part of this puzzle is the way the design team operates to integrate sustainability.

Sustainable design is nothing new, but it is evolving and moving from a specialist field to the mainstream. As the impact of construction upon the environment heightens in public awareness and expectations of performance increases, many architects are adopting changes to their approach to the design process – not just to secure work but also to fulfil client expectations for the completed project. Sustainable refurbishment can be seen as an ambiguous concept dependent on many factors, including, the adaptability of the existing construction, future intended uses and servicing requirements in addition to the expectations of the client and end users.

Increasingly, buildings are shaped by demanding government standards, such as Part L of the British Building Regulations. It is the role of the design team not only to be aware of these changing criteria and possible future developments, but to work with them to create a scheme which satisfies the expectations of the client and allows for future adaptation in use. Establishing clear processes for continual improvement and design development, including undertaking sustainability reviews, and setting objectives and targets throughout the planning, design and construction phases, will be beneficial to not only the project directly but equally to subsequent schemes, where lessons learned by the team, client and agents can be continuously applied and improved upon.

Standard design processes are often based on achieving targets defined by a recognised assessment method such as the Building Research Establishment's Environmental Assessment Method (BREEAM) in the UK or the US Green Building Council-developed Leadership in Energy and Environmental Design (LEED) system, which is not only regularly applied to projects in North America but frequently discussed in the context of commercial projects elsewhere. Often these assessment methods are a useful tool with which to benchmark the performance of the building (similar to the EPC – Energy Performance Certificate, broadly used to determine energy performance) and are more frequently becoming a condition of planning consent. Commercial agents can also capitalise on growing public attention to sustainable credentials by communicating the building's achievements once completed.

However, these assessment models, no matter how effective in asserting positive changes within the building design, are time-consuming processes and as such are undertaken at only two or three stages within the design and construction process. The infrequency of this review process may mean that the design team does not fully engage in understanding the potential for further sustainable improvements beyond targeted achievements.

Commercial projects are frequently put on hold due to developer priorities or for investment reasons. Often, when these schemes are resumed from an earlier stage of development, building standards have developed, causing a reactionary chain of design changes. By adopting new practices and raising the sustainability awareness of staff, joint venture partners and manufacturers, a more dynamic, proactive and forward-thinking design process can be developed to enable buildings to pre-empt or flexibly adapt to future performance requirements.

STIMULATING THE DESIGN TEAM

It cannot be the place of only one individual within a project team to represent the 'green agenda' of a scheme. Project teams should be encouraged to take part in sustainability reviews as an integral part of the design and development process, which will serve to reinforce and sustain the appeal and value of the work as a whole.

Knowledge can be disseminated and shared within and between teams in a variety of different ways, for example:

- project-related research and development;
- group forums;
- mentoring by individuals with more developed sustainability credentials;
- a comprehensive sustainability intranet database (e.g. Latham and Swenarton, 2007);

or

- a conventional practice library.

The usefulness and momentum provided by a resource, such as a group of people whose previous sustainable experience is tangible and accessible, is highly significant in supporting other project teams through the design, evaluation and development of a project. In order to encourage and stimulate understanding of sustainability issues it is also important to demystify the terminology used. Acronyms and metrics are used in abundance around 'sustainable' schemes and often these are misunderstood and therefore either become irrelevant or trivial. Clear, transparent communication of the cause and effect of sustainable options is key, both within the team and when client-facing.

AHMM SUSTAINABILITY TOOLKIT

The Green Group, formed by Allford Hall Monaghan Morris Architects (AHMM) in 2005 is an example of an established, active and well-supported resource of multidisciplinary people within a practice, which allows the company to direct further development in the most effective manner. Rather than focusing on isolated projects, the group works across the practice and has grown to include a significant number of architects working at every level and in every sector in order to identify good practice and define and achieve key standards.

AHMM have developed a 'Toolkit', an assessment tool that does not define the extent of the approach to sustainability but instead helps to assess performance and further assist in the design of well-considered and appropriate buildings and places.

The Toolkit is formed of four components (Figure 6.1):

- guidance manual;
- design prompt spreadsheet;
- assessment guidance;
- a graphic output.

Sustainable aspects and environmental performance in twelve key areas relating to carbon, people and ecology are covered:

1 operational energy;
2 transport;
3 waste;
4 renewable energy;
5 climatic design;
6 embodied energy;
7 amenity/community;
8 occupants;
9 materials;
10 safety/management/accessibility;
11 water;
12 biodiversity.

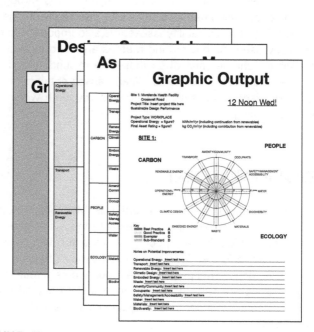

Figure 6.1 The AHMM Toolkit components
Source: AHMM

AHMM Toolkit sessions are run at every major design stage of a project in a workshop format to explore opportunities and challenges with the results illustrated on a simple diagram, the Toolkit 'rose', which enables comparison and review. Results are reviewed internally during formal work stage reviews and during post-completion feedback sessions. The Toolkit rose highlights areas worthy of further consideration for the individual project but equally for ongoing collective performance. The rose, which simplifies performance to a graphic traffic light system; dark green (good) to red (bad) can be used by a client or end user, among others, to better understand the project's sustainable performance (see Figure 6.2).

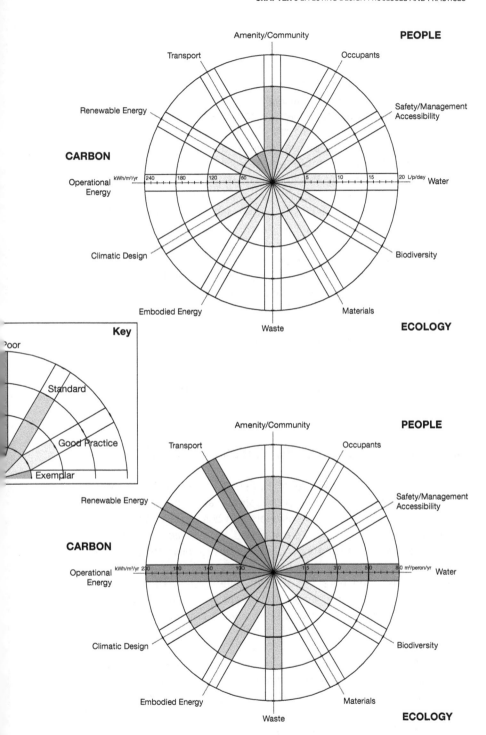

ure 6.2 High Achiever (Weston Street) vs. Room for Improvement (Maywood Park)

urce: AHMM

The entire design team attend the Toolkit review or, in the case of very large schemes, key team members for each significant aspect of the scheme. It is the responsibility of the team to present a brief overview of the scheme and by using the twelve key headlines, positive aspects of the scheme and areas where it is struggling to perform can be identified. The review is chaired by the sustainability consultant who originally helped to develop the Toolkit, along with a member of the AHMM green group. This balance of 'Sustainability professional' and 'Sustainability enthusiastic' ensures a broad resource of precedent projects, strategies and innovative product development ideas. The rose, along with its summary outcomes, can be circulated to external consultants (mechanical and electrical (M&E)/ structural, etc.) in order to instigate further design developments. Every so often a sustainability benefit can be clearly identified as a clear exchange of one design feature or specification for an alternative. However, more frequently, other members of the project team such as the quantity surveyor may need to engage in cross-examination of a proposal in order to quantify its benefit, either sustainably or commercially, to the project. This is where a project team becomes reliant on healthy, engaged working relationships to develop the scheme dynamically.

As Toolkit reviews are performed frequently, a deadline for resolving issues identified in previous reviews is apparent. A forthcoming Toolkit review date ensures that issues identified are continually developed and usefully discussed at forums where all project team members are present.

HOW SUSTAINABILITY IS INTEGRATED INTO THE DESIGN PROCESS

The original UK Royal Institute of British Architects (RIBA) work stage arrangement treated each project in isolation as a linear process. This is reflected in the way many professionals and businesses operate and how relationships within the industry are arranged. With the advent of RIBA's 2013 Plan of Work we are seeing a concerted effort to move towards a cyclical review, design and development process. This cyclical assessment method moves towards a more holistic or systems-thinking approach, acknowledging that projects don't exist in isolation, allowing for reciprocal learning from each scheme. This approach integrates perfectly with the Toolkit methodology, where lessons learnt on one project help to broadly inform subsequent projects (Figure 6.3).

AHMM's own office consists of an additional 840 m^2 lightweight floor built above the open-plan refurbishment of an existing warehouse building which was completed in 2013. Having identified beneficial characteristics such as the existing building's location, orientation and construction typology (solid masonry), the design proposal was reviewed in a Toolkit, encouraging an already enthusiastic client to push the brief further in order to achieve BREEAM Outstanding – a first for a commercial refurbishment

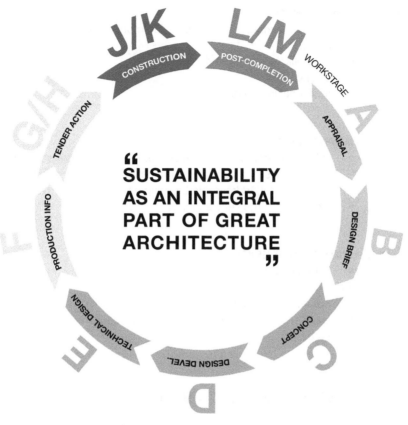

Figure 6.3 Rotary RIBA stages

Source: RIBA

for both the client and practice. Having run Toolkit reviews on buildings completed in the past, the team was better equipped to understand how lessons learnt since could be best applied. Additionally, since completing the scheme, monitoring equipment has been installed to constantly assess the performance of the new construction and document the findings for use on future schemes.

By monitoring projects and their success/failures, AHMM have seen that it is not always necessary to roll out the carpet on every refurbishment or reuse scheme. It can be argued that long-term value is not to be found in creating a typologically defined single-use space with tightly constrained servicing strategies, but rather, architecture responsive to change. This aspect of sustainability, not traditionally considered, is becoming a more popular concept to the market – ultimately a hackable unit, one that can be chopped apart and altered in order to facilitate the changing requirements of its occupants. Three incremental stages can be useful to understand what is required to substantially improve the building whilst allowing for future adaptation:

Stage 1 – passive measures

- window upgrades to improve thermal and acoustic performance;
- controllable background ventilation;
- solar control;
- additional insulation, etc.

Stage 2 – active measures

- lighting systems and controls, e.g. passive infrared (PIR – motion-sensitive) controls, photocells switches or a fully integrated Digital Addressable Lighting Interface (DALI) system;
- night-time purge ventilation, etc.

Stage 3 – comprehensive measures

- hybrid cooling/heating;
- high-efficiency plant;
- hot and cold water thermal loop;
- local heat-exchanger/localised cooling units for high-capacity areas, etc.

USE OF THE TOOLKIT ROSE OUTSIDE THE DESIGN TEAM

A critical aspect to developing a sustainable commercial building successfully is to engage not only the client in early dialogue about available options but also to communicate to the agents the aspirations of the scheme and how it could work to its full advantages. By using the Toolkit rose, or a similar achievement level diagram, it can quickly be identified if there is potential to excel beyond traditional standards or 'good practice' in order to achieve a better market return.

Irrespective of which methods are undertaken, benefits will only be realised if results are analysed, considered and used to inform future development discussions from inception and briefing onwards.

There are also often good opportunities to engage clients in discussions promoting smarter ways of building. Discussions motivated by both commercial projects and academic research have included:

- ultra-low operational energy design;
- upgrading the existing building stock for increased return;
- developing environmental mock-ups;
- improving indoor air quality.

FEEDBACK LOOPS AND THE FUTURE

Allowing for adaptation is important in order to avoid an environment which becomes redundant of purpose, or unmanageable. User, tenant and agent expectations have a significant impact on the parameters of commercial refurbishment briefs and subsequent modifications. As such, looking forward is key to developing priorities within a commercial refurbishment. By making efforts to identify future scenarios we can inform both internal development and equally assist clients in commissioning suitable design outcomes.

With mobile technologies developing rapidly, never has the will of the worker or consumer been so critically important. Social media tools such as Twitter and Instagram are carrying the qualities of the work zone into a dimension that is still to be understood fully, with reviews and documentation of spatial and environmental experiences becoming immediate. A digital interface may well be the next step to integrating feedback to a commercial agent or building manager, but can we as designers also utilise this dialogue in order to develop and 'feed in' to a cyclical design process?

REFERENCE

Latham, I. and Swenarton, M. (2007) *Feilden Clegg Bradley: The Environmental Handbook*, London: Right Angle.

7
RETROFITTING FOR COMFORT AND INDOOR ENVIRONMENTAL QUALITY

Nick Baker

The motivation for refurbishment of and retrofit to existing buildings can be driven by a number of objectives. It could be simply to improve the energy performance, and as we now say, sustainability of the building. But if this is the case, it should always carry the constraint that the comfort of the occupants is not compromised. On the other hand, in buildings that currently deliver poor environmental quality – e.g. summer overheating, poor air quality, etc. – common defects in existing buildings, the main objective may be to improve comfort. To achieve this without incurring greater energy costs may be a secondary objective.

This chapter looks at the opportunities for achieving improved indoor environmental quality whilst carrying out measures that are known to have energy performance benefits. It is aimed at an integrated approach, where the energy performance and environmental implications are considered simultaneously.

For a rather obvious example, an over-glazed lightweight building currently overheats in summer causing serious comfort problems. Installing a state-of-the-art air-conditioning system, with a guaranteed upper temperature limit, could achieve an improvement in comfort. However, there would remain local control problems: occupants sitting near windows would still receive direct radiation and, of course, the building's energy consumption would increase significantly.

Alternatively, the facade could be remodelled, reducing glazing area, adding controllable shading devices and easily modulated openable windows. This would greatly reduce the source of the overheating and provide a personal and intuitive means of controlling fresh air and glare. It is likely that the overall satisfaction of the occupants would be greater in the second case, and almost certain that the energy consumption and maintenance costs would be much less.

ENVIRONMENTAL COMFORT

Before examining sustainable retrofit measures in detail, we will briefly review the basic principles of environmental comfort. Of the various environmental parameters that affect our comfort, temperature is probably the most critical, and one we spend most effort to control. This is probably because of the range of temperatures that the outside environment covers, on a daily and seasonal basis, often includes conditions that are not only very uncomfortable but possibly life-threatening.

Thermal comfort theory

For the latter part of the last century, the conventional wisdom was that thermal neutrality should be the target. It seemed very reasonable that people are comfortable when they are neither hot, nor cold. Experiments were carried out in climate chambers by Fanger (1970) to establish neutral temperature, and the results were used to establish international temperature standards. It was upon the perceived need to achieve these, in both the workplace and at home, that the air-conditioning industry was built.

The four components of heat loss from the body, in moderate climate conditions, are shown in Figure 7.1. Conventional comfort theory assumes the comfort temperature is synonymous with neutral temperature, i.e. to be where the total heat loss exactly balances the metabolic heat gain, on an instantaneous basis. Because neutral temperatures were established in closely controlled conditions, the standards did not recognise adaptive behaviour and psychological affects, as discussed below.

Note that only 20 per cent is by evaporation. Of this about 13 per cent is by respiration and base-level diffusive evaporation. Only 7 per cent is by evaporation from moist or wet skin, and only this part is sensitive to the ambient humidity, provided it is between about 30 and 70 per cent relative humidity (RH).

In spite of this, conventional air conditioning has provided close RH control, often as close as 65 per cent +/- 10 per cent. This requires a significant amount of energy, which is completely wasted as it has no detectable effect on comfort. It is interesting that dryness of skin and eyes is often reported as a symptom of Sick Building Syndrome (discussed later), implying

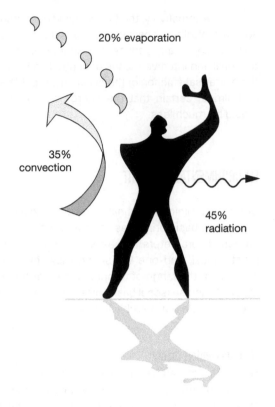

Figure 7.1 The four components of heat loss from the body. Conventional comfort theory seeks to balance total heat loss with metabolic heat gain at all times

Source: N. Baker

20% evaporation

35% convection

45% radiation

that in spite of close RH control targets, air-conditioned buildings often deliver much too low humidities, probably due to poor maintenance and control.

It is only when in warm humid conditions, often met in the humid tropics, that humidity control, in this case dehumidification, will be justified. In these circumstances, due to high air temperature, evaporative cooling from moist skin and clothing becomes a dominant heat loss mechanism, and lowering humidity will increase comfort both from a temperature perception, and the reduction of skin wettedness.

Adaptive comfort theory

The focus on close temperature and humidity control dominated the air-conditioning design and specification for half a century. However, it was pointed out by Humphreys (1978) as early as 1978, that when comfort surveys are carried out in the home and workplace instead of comfort chambers, people report comfort over a much wider range than predicted from Fanger's heat balance theory. Humphreys also noticed that the degree to which the occupants can take adaptive action greatly influenced their satisfaction. This is illustrated in Figure 7.2, which shows a hypothetical stress response of an occupant to swings in an environmental parameter (such as temperature), away from neutral. It goes some way to explain why

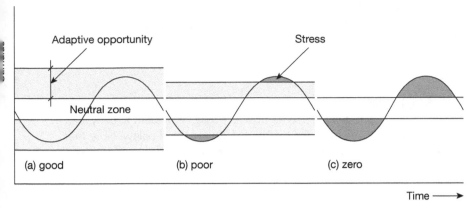

Figure 7.2 Human stress response to swings in an environmental stimulus (such as temperature) and its relation to the presence of adaptive opportunity

Source: N. Baker

there appeared to be such a discrepancy between Fanger's findings in the climate chamber, where all *adaptive opportunity* is removed, and in the real world where in most cases some adaptive opportunity exists. Later work by Guedes and Matias (2009) showed the *perceived* presence of adaptive opportunity to have a beneficial effect *even when the opportunity was not actually taken*. This suggests that there is a strong psychological component to environmental satisfaction.

Adaptive action, and the resultant benefits, extend to the control of mechanical systems. Unfortunately, many buildings are controlled by centralised management systems, which discourage or even prevent any interaction from the occupant. In other cases, controls are difficult to understand and access. Intuitive control, where the occupant carries out an action which is obvious, easy and has a rapidly responding effect (analogous to opening a window), should always be sought.

This is highly significant to the matter in hand. Only the use of energy-consuming air conditioning can guarantee close temperature standards. However, the knowledge of adaptive comfort theory, and the recognition of building elements and design that provides adaptive opportunity, can produce alternative sustainable solutions.

There are parallels in other environmental variables too. In lighting, many designers have given up on daylight as a working illumination, claiming that it was too variable and unpredictable. High levels of uniform artificial lighting became the norm, and daylight was more to do with building aesthetics than function. This was largely due to a similar misunderstanding – the notion of uniformity and neutrality. Studies by Papairi (2004) have shown that occupants prefer daylight in spite of poor illuminance geometry and intensity, provided they have some adaptive opportunity to mitigate its negative effects. It was also shown that far from ideal lighting conditions were acceptable if they we seen to be due to 'natural causes'. This phenomenon can be observed when the occupant of a room lit by the failing evening daylight, carries on a visual task such as reading, at illuminances as low as 50 lux, 17 per cent of the recommended 'standard' value, without switching the lights on.

Adaptive opportunities

Typical adaptive opportunities, some of which may be present already, and some of which may be provided by the retrofit, are listed below. Note that some relate to the personal environment – e.g. dress code – whilst others relate more to the room environment – e.g. openable windows.
Positive adaptive attributes:

- relaxed dress code;
- occupant mobility;
- access to hot/cold drink;
- openable windows;
- adjustable shading/blinds;
- desk fan or locally controlled ceiling fan;
- local heating/cooling controls;
- workstation/furniture flexibility;
- shallow plan (minimising distance from window);
- cellular rooms (reduces mutual disturbance);
- daylighting with task lighting back-up;
- good views (external and internal);
- transition spaces (e.g. balconies, atria, etc.);
- good access to outdoor areas.

Negative adaptive attributes:

- uniformity of the physical environment (temperature, lighting, colour);
- deep plan, reduced access to perimeter;
- dense occupation with restricted workstation options;
- sealed windows;
- views obstructed by fixed shading devices and screens;
- central control of mechanical services.

What does all this mean for sustainable retrofit? It is good news. It means that many retrofitting measures that are consistent with good energy performance (such as shading, natural ventilation, provision of daylighting) are also con-ducive to improved comfort, provided three basic principles are observed:

1 The retrofit measures, together with the existing characteristics of the building, limit the range of environmental conditions (e.g. avoiding extreme temperature swings) by passive means.

2 The retrofit measures should provide sufficient adaptive opportunities together with those already existing to cope with the environmental range.

3 Control systems should anticipate and encourage occupant intervention.

There is also another positive aspect when a number of environmental issues are addressed, that Leaman and Bordass (2007) refer to as the 'forgiveness factor'.[1] Based upon data collected in many building occupant surveys, they have found that although individual environmental parameters (e.g. temperature, air quality, lighting, etc.) may score only moderately well, when it comes to reporting the overall satisfaction with the building, occupants report a higher score, in certain types of building. These buildings are nearly always those where control is visible and intuitive, and are predominantly passive, rather than highly mechanically serviced.

This is consistent with the earlier assertion that occupants are more tolerant of non-neutral conditions when they perceive them to be due to natural causes, and we can add to that – when they understand the means by which they could mitigate them.

Healthy planning

There is a popular perception that efficient planning, especially in the workplace, means minimising the need to move away from the home base. This is particularly so in office buildings, where for decades the 'ideal' office has been one where circulation distances are minimised, and technology, such as email, has made face-to-face meetings unnecessary.

However, recent studies by Rassia *et al.* (2010) have shown that on average, occupants consume more energy moving around their workplace than at home in their leisure time. This means that workplace activity provides important exercise. If the current trend to minimise energy expenditure in the workplace could be reversed, it could help to improve fitness and reduce obesity. For example, simply standing up and walking 5 metres to a 'waste centre' uses seventeen times more metabolic energy than leaning forward and throwing a paper into the desk-side waste bin; using the stairs uses twenty-four times more metabolic energy than using the lift.

Rassia *et al.* identified the need for 'rewards' for actions that required some effort. So often it is the lift that is the luxury element and the stairs are cold and smelling of cleaning fluid (if you are lucky). But if the staircase offered fine views across the landscape, or contained art objects, or provided music, maybe people would prefer to use the stairs.

This principle could be considered when remodelling and retrofitting, and is consistent with a more holistic approach to occupant comfort and well-being.

Contact with nature

It is not uncommon to explain the role of the building as a moderator between the internal and the external environment. This is partly achieved passively, mainly by the fabric of the building, and partly by the mechanical services. Even in an urban situation, the external environment is dominated by the natural climate – e.g. sunlight, rain, wind, etc. – and may include vegetation and landscape. Thus, since the internal conditions are being to some extent driven by the natural world outside, it is important that some kind of contact to the outside can be made by the occupant. This could be purely visual, and could influence the choice of glazing design and specification, or could involve actual contact with the outside – e.g. access to courtyards, roof gardens, terraces, unheated atria, etc.

There is a growing opinion that this need relates to our genetic background. It is only relatively recently in genetic terms, (perhaps five generations), that humans have moved indoors where they now spend 95 per cent of their time. Closely related to this, there is also the belief that the exposure of occupants to high levels of daylight is essential for so-called circadian entrainment, that is the synchronising of body rhythms with the day/night cycle.

The relevance to sustainable retrofit is, first, that the existing building may already have positive attributes in this respect, which have been compromised by successive minor alterations and modifications. For example, it is not unusual for clear glazing to be treated with obscuring films, or for doors to outside spaces and terraces to be locked for security reasons. The same applies to adaptive opportunities listed earlier – windows may have been sealed, and blinds and shades removed, accessible heating controls disabled.

The considerations above point to a holistic approach to sustainable retrofit, where the whole is greater than the sum of the parts. In other words, concentrating upon a single environmental factor, such as overheating, for example, by providing air conditioning, could be less effective overall than a range of environmental measures, even if in themselves they could not achieve such close temperature control as air conditioning.

THE BUILDING FABRIC AND COMPONENTS

Thermal insulation

The motivation for thermal insulation is usually heating energy conservation. Buildings from the 1950 to 1980 period were very poorly insulated, and with the reduction of internal gains due to modern equipment and lighting, auxiliary heating is large. Significant savings in auxiliary heating energy can be made by upgrading the envelope insulation.

In terms of comfort, improvements to the envelope insulation can reduce under-heating in winter, in particular local under-heating. For example, a previous alteration may have partitioned a part of the floor plan in such a way that it included a disproportionate amount of external envelope to the installed heat input. By increasing the external envelope thermal resistance, temperature differences across internal partitions, and within spaces, will be significantly reduced.

Envelope insulation is sometimes perceived to have a deleterious effect on summer comfort due to overheating. This view has largely been prompted by simulation results, where for a given internal gain and ventilation rate, increasing the thermal insulation will indeed increase the internal temperature, for some periods of the warm season. However, if there are openable windows, it is far better to create heat loss by ventilation rather than by fabric loss, since this mode is controllable and does not have to operate during the heating season. Furthermore, other actions of the retrofit will probably reduce internal gains due to reduction in lighting loads and improved efficiency of equipment.

Envelope insulation may improve summer comfort in some circumstances. In poorly insulated buildings, overheating is often experienced in top floors due to poor roof insulation. Overheating could also be caused by a poorly insulated wall when exposed to low-angle sun. In both cases, increasing the insulation will reduce transmitted solar gain and reduce overheating. Increasing the reflectance will also reduce overheating by reflecting the solar energy away from the surface.

Thermal mass

Thermal mass has an important role in stabilising temperature and, thus, in a building without mechanical cooling, limiting peak temperatures. This has obvious implications for comfort, and makes the adoption of passive features as an alternative to air conditioning more viable.

One technical issue that is relevant to both design and comfort is the coupling of the thermal mass with the occupant, as well the source of heat gain. Figure 7.3 shows how the thermal mass in the walls and ceiling are coupled to the heat source (the sun patch), and to the occupant, by both convection and radiation. Under normal conditions, heat loss from the body by radiation is around 45 per cent, and is greater than convective and evaporative loss. Thus the radiative coupling, by direct line of sight to the mass from the body, is a positive design feature, although indirect coupling by convection is still important.

In retrofit projects, it may be possible to expose thermal mass that has hitherto been isolated by finishes, thereby meeting the criteria above. Typically, suspended ceilings and raised floors could be removed (or partially removed) to expose massive floor slabs. This could have negative effects on acoustics and noise control, which will be discussed later. Exposed massive ceilings are very effective as they are usually unobstructed allowing radiative coupling with the occupants, and convective coupling to room air.

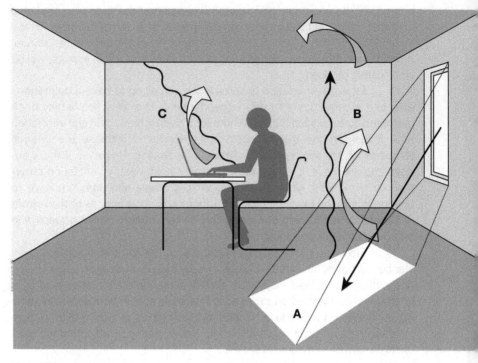

Figure 7.3 Coupling between thermal mass, sources of heat gain, and the occupants is essential for thermal mass to be effective. (A) solar radiation converted to heat at sun patch, (B) heat lost by radiation and convection to thermal mass in soffit, (C) heat lost from occupant to thermal mass

Source: N. Baker

Glazing

For highly glazed buildings (e.g. main facades with 50 per cent glazing or more), with single-glazing, losses through the glass are likely to be the largest component of fabric heat loss. Moreover, older framing systems are often leaky to air infiltration and, due to cold-bridging and their 'fin effect', can lead to a U-value significantly larger than the glass itself. Thus glazing, together with its support structure, is a common object for retrofit. Typically the facade will be reglazed using double- or triple-glazed units in thermal break framing, which can easily reduce the overall U-value from around 6.0 to 1.0 W/m²K.

This has several impacts on thermal comfort. In particular, in cold conditions, the mean radiant temperature near to the glass is raised to much nearer room temperature, as indicated in Figure 7.4. This means that it does not require so much compensation from perimeter heat emitters. The cold inner surface of single-glazing also causes downdrafts, which can lead to strong vertical temperature gradients, with discomfort caused by cold feet and legs.

In summer conditions, the extra insulation value of the glass will only be of significance when there is a significant temperature difference between inside and outside. If naturally ventilated in a temperate climate,

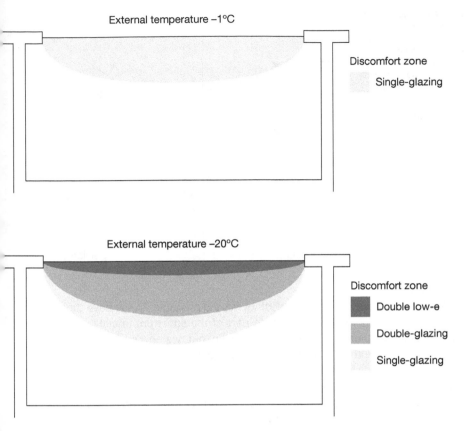

Figure 7.4 The impact of double-glazing on mean radiant temperature and comfort close to window
Source: N. Baker

with high air change rates, these differences will generally be small. The main impact will be a reduction of mean radiant temperature where the glazing is receiving direct radiation, due to the lower transmissivity of the multiple panes and the low-emissivity (low-e) coatings. However, this benefit will be modest and should not be seen as an alternative to shading devices.

Tinted and reflective glazing

Many older buildings have already been fitted, or retrofitted, with tinted or reflective glass or films. Figure 7.5 shows the mechanism of transmission and reflection for different glass types and clearly explains why tinted single-glazing is so unsuccessful. Reflective glass performs better, but consideration has to be given to the possibility of reflection of solar energy into other adjacent buildings.

Tinted or reflective glazing reduces the luminance of the outdoor scene by the value of the glazing transmission, and for transmissivities of less than 25 per cent (quite common in older installations) the reduced brightness might be significant to occupant well-being in winter.

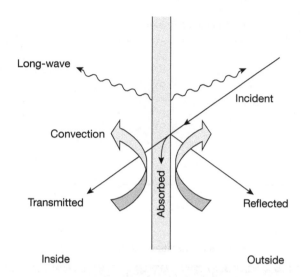

Figure 7.5 The mechanism of transmission, reflection and absorption for glass. Absorbed radiation heats up the glass. This heat is lost by long-wave radiation and convection to the outside and to the room

Source: N. Baker

Long-wave

Incident

Convection

Transmitted

Absorbed

Reflected

Inside

Outside

Selective glazing

Modern glazing materials include selective glazing products. These have the ability to absorb and reflect the invisible part of the spectrum (mainly near infrared) that contributes to unwanted solar gain, but has no visual function. This improves the luminous efficacy of the daylight.

Glazing area

Many environmental problems are caused in part by over-glazed unshaded facades. Apart from the energy implications, these often are a direct cause of discomfort due to local thermal effects (as described above) and glare. In most cases, the motivation behind the large area of glazing is stylistic rather than functional; many studies show that for a side-lit room, areas of glazing greater than about 40 per cent of the facade have little functional benefit. This means that in a building retrofit, a reduction of glazing area and its replacement with well-insulated opaque panels should be considered. This brings comfort benefits, and also internal space use benefits. And, if the glazing requires shading, it means that the area of this costly element is reduced too.

If reduction of glazing area is being considered, three principles should be applied:

1   The upper part of the glazing contributes most to the penetration of daylight to the back of the room – the amount of reduction should take account of the depth of the room.
2   Mid-level glazing provides important distant (horizon) views – high sills cause irritating obstruction.
3   Below work-plane glazing only contributes to daylighting via external reflection from the ground and internal reflection from the ceiling, and provides near views, dependent on the floor level above the ground.

Increasing glazing may be appropriate in some cases. Deep-plan single-storey or top-floor rooms can be daylit over their whole area by roof lights. The geometry of the new roof lights and their shading should be taken into account in relation to the function of the space beneath. In many buildings, existing roof lights may have been obstructed under the misguided belief that it will solve an overheating problem. The real cause is often poor roof insulation, and the roof lights should be reinstated, with appropriate shading, as part of the upgrading of the roof.

Increasing glazing in side-lit rooms may be appropriate to improve view or daylighting. A particular case of the latter is where a suspended ceiling has been removed, making a high level of the opaque wall available for glazing; this will improve daylight penetration to the back of the room, as in Figure 7.6.

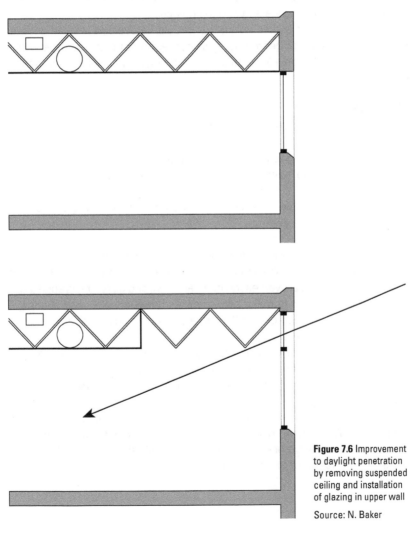

Figure 7.6 Improvement to daylight penetration by removing suspended ceiling and installation of glazing in upper wall

Source: N. Baker

An overall criterion when changing glazing areas is that for buildings with specific workstations such as offices, every workstation should have a distant outdoor view, and a daylight factor of at least 1.5 per cent.

Shading

Shading is an integral part of good daylighting design, particularly on facades that are insolated at some time of the year. This is largely because direct sunlight is between five and ten times the brightness of the diffuse sky, but due to weather, it is not predictable.

Fixed shading

Shading may be fixed, relying on its geometry to selectively obscure parts of the sky through which the sun passes. These devices could be in the form of fixed louvres, overhangs and/or deep reveals.

One type of fixed shading that should be avoided is fixed grids or screens. These are not geometrically selective, do not improve the thermal performance of the envelope (unlike reducing glazing area) and seriously disrupt the view.

Adjustable shading

Shading can be adjustable in three ways:

1 variable geometry such as in louvres or fins;
2 variable transmission – also achieved by closable louvres or fins;
3 deployable – i.e. can be deployed in or removed completely from the aperture.

Shading devices may be located inside, outside or mid-pane.

Well-designed shading has a positive effect on comfort, reducing room overheating, local overheating from radiation, and glare. If it is adjustable, it allows the occupant to make compromises between view and shading if necessary, and provides one of the more significant adaptive opportunities that promotes comfort satisfaction.

In choosing a type of shading for retrofit, its impact on natural ventilation must be considered. For instance, a translucent roller blind will completely obstruct airflow, whereas louvres will allow it to a greater or lesser extent.

The properties of shading devices and their suitability are summarised in Table 7.1. An overall criterion is that for buildings with specific workstations such as offices, every workstation should have a distant view, and a daylight factor of at least 1.5 per cent.

Shading type	Orientation	View	Nat vent (in limiting conditions)	Daylight (in limiting conditions)	Seasonal response	Modulation	Notes
Overhangs Fixed Retract	180 +/− 30 180 +/− 30	Good Good	Good Good	Medium Good	Medium Good	None Good	e.g. Canvas awnings + adjustable geometry
Louvres Fixed Adjust Retract. Retract.+ adjust	180 +/− 30 All 180 +/− 30 All	Med – poor Med – poor Med/good Med/good	Good Good Good Good	Medium Good Good Good	Medium Good Good Good	None Good Medium Good	View influenced by blade module size and geom. 'Good' applies to when retracted
Fins (vertical) Fixed Adjust	90, 270 +/− 20 90, 270 +/− 45	Med – poor Med	Good Good	Medium Good	Medium Good	None Good	View influenced by blade module size and geom.
Blinds Retract	All	Poor/good	Poor	Good	Good	Medium	'Good' applies to when retracted
Perforated screens Fixed	All	Poor	Med – poor	Poor	Poor	None	Not recommended

Source: N. Baker

Notes to accompany Table 7.1:

i   Orientation all implies that performance is not orientation-sensitive, although in general, there would be no demand for shading on facades orientated towards the poles (N or S) +/−45°.

ii Adjustable louvres and fins will have poor view performance and poor natural ventilation performance, when completely closed.

iii Natural ventilation limiting conditions refers to situations requiring the shading to be deployed, that is when there is an overheating risk. Good implies that natural ventilation is not impeded.

iv Daylight limiting conditions refers to minimal daylight availability when there would be no requirement for shading. Good implies that the shading device causes no reduction in daylight transmission in these conditions.

v   Distribution refers to the ability of the shading device to improve daylight distribution, thereby lowering the total daylight energy (solar radiation) required.

vi Seasonal response refers to the ability of the shading device to respond to different sun angles at different seasons.

NATURAL VENTILATION

The existing building will already rely on either natural ventilation (apart from toilets, kitchens, etc.), mechanical ventilation or combined ventilation, heating and cooling (air conditioning). Some older buildings will have had various levels of air conditioning installed during the building's lifetime, often in response to poor environmental standards due to inherent design deficiencies. These deficiencies typically include excessive solar gain and heat loss through over-glazed leaky facades, poor indoor air distribution and poor control of high pollution areas. If these deficiencies can be removed or mitigated, it should be an objective of the retrofit to return the building to natural ventilation.

In older buildings, infiltration (uncontrolled ventilation due to a leaky envelope) may be making a significant contribution to air quality, and this could be compromised if the upgrade to the fabric includes the reduction of infiltration. This must be compensated for by providing purpose-made controllable fresh air sources, either driven by natural or mechanical means. A guiding principle is 'build tight – ventilate right', meaning that ventilation should be demand led, and if naturally driven, the wide variation in the driving forces of wind and stack effect compensated for by controls.

The three functions of ventilation
It is important to separate out the functions of natural ventilation. These are:

1 provision of a minimum air quality (dilution and removal of pollutants);
2 the removal of unwanted heat gains (to avoid overheating);
3 to provide air movement for direct physiological cooling.

When ambient temperatures are low, only the first function is required, and the relatively low rates can easily be provided by wind or stack effect. At medium ambient temperatures, in the absence of wind, in densely occupied areas, natural ventilation may be unable to provide sufficient fresh air, and mechanical back-up should be considered. This can be a very satisfactory solution in a well-controlled predominantly naturally ventilated building, where the mechanical back-up is controlled by detecting CO_2 levels, a good indicator of indoor air quality. The mechanical equipment can be located in elements that are part of the passive system, such as stacks or roof ventilators.

For the second and third functions, high air change rates are required, necessitating large openable areas. These are normally provided by windows, thus combining the three functions of daylight, view and fresh air into one element, reinforcing intuitive control by the occupant.

However, there may be some circumstances where due to ambient noise or pollution, opaque ventilation components (flaps, dampers, etc.)

incorporating sound attenuation and or filtration are required. In terms of ambient noise control, the design of the window can have considerable influence as illustrated in Figure 7.7, and this should be considered when specifying retrofit windows.

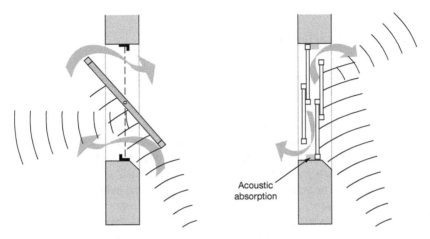

Acoustic absorption

Figure 7.7 Window design for enhanced noise reduction for natural ventilation in noisy locations

Source: N. Baker

Indoor air quality

The provision of minimum air quality, as defined above, is normally concerned with removal and dilution of CO_2, water vapour and bio-odours. Bio-odours – produced by the occupants – are in themselves harmless, but act as a warning that air quality due to occupation may be insufficient. There may also be other pollutants produced by activities associated with occupation, particularly in manufacturing processes, cooking and even some office equipment (e.g. ozone produced by photocopiers). These pollutants may or may not produce odours detectable by the occupant. Toxic pollutants with no odour are the most dangerous (e.g. carbon monoxide), and where there is risk of these, special detectors must be installed. Some will be dealt with by specialist local extract ventilation.

Of particular relevance to a newly occupied building after retrofit, is the outgassing of new materials. This is mainly due to volatile organic compounds (VOXs) present in adhesives, paints and some recently manufactured materials such as laminates and textiles. Again, these may or may not produce odour, and generally will not be highly toxic since this will be covered by regulation. But they may cause discomfort to occupants, and possible more serious and unpredictable symptoms in certain individuals due to allergic reaction.

Outgassing usually takes place over a short period relative to the life of the building, and may have progressed sufficiently before full occupation. There is some evidence that the process takes place quicker at higher temperatures, and this could be arranged in the pre-occupied period. Once the building is occupied, the effects of residual outgassing can be mitigated by a higher than normal ventilation rate, for the first few weeks. The building could also be heated during the unoccupied period.

NOISE AND ROOM ACOUSTICS

The comfort and well-being of occupants can be affected by the acoustic properties of a building in four ways:

1   The ingress of noise from outside through natural ventilation.
2   The noise level inside the building due to sounds generated within the building. This may be due to the transmission of noise from another part of the building (including structure-borne sound) or the reverberation of noise generated within the space.
3   Poor speech intelligibility due to long reverberation time.
4   Loss of privacy due to the transmission of sound in the building.

All of these may be influenced by the specification of the retrofit.

The control of external noise can be achieved by window design. It can also be controlled by external treatments to the site landscape. This can include the provision of barriers or in some cases the removal of reflecting surfaces. It is quite conceivable that noise control measures carried out as part of site landscaping could enable natural ventilation to be viable, where otherwise a sealed air-conditioned building would be the only solution.

Conflicts with natural ventilation

Unfortunately, the provision of flow paths for natural ventilation within the building often link rooms and permit sound to travel throughout the building. Where this is achieved through casual openings (doors, internal windows, etc.) it is difficult to control and may have to necessitate a compromise by the occupants between thermal and acoustic comfort. In some cases circulation spaces could be used or, in more engineered solutions, bypass ducts can be included to achieve cross-ventilation in double-banked rooms.

Loss of privacy is closely related to noise transmission. The main difference is that intelligible speech needs to contain high frequencies, which means that reverberant sound (see below) is less likely to cause a problem. Privacy problems are commonly caused by flanking transmission or by first reflections from ceilings. Curved reflective surfaces may cause some focusing effects and amplify reflected sound with unexpected results.

Reverberation and exposure of thermal mass

Due to the ear's wide-ranging sensitivity to sound energy (a range of 10^{12}), audible sound will persist in a room until the energy is absorbed at the room surfaces to a very low level. This time, the reverberation time (RT) is typically in the order of one or two seconds for an acoustically 'live' room, or as short as 1/10 second for an absorbent room.[2]

Reverberation has two main effects on the room acoustic. First, long RTs cause a build-up of reverberant sound that includes unwanted noise – either generated within the space, or outside – and reduces speech intelligibility by overlapping the high frequency syllable sounds due to multiple reflections remaining audible.

The absorption being referred to here generally takes place when the sound is reflected off the walls, floor and ceiling of the room; highly absorptive finishes involve fibrous and porous layers and/or flexible panels. Unfortunately these finishes also thermally isolate the room from any massive structural materials, and thus good acoustic performance is in direct conflict with the provision of thermal mass for temperature control.

There are measures that can be taken to affect a compromise. Exposed floor slab soffits, particularly effective for thermal transfer, can be provided with hanging acoustic absorbers which still allow convective and some radiative transfer of heat, as shown in Figure 7.8. Soft furnishing and partitioning systems can also provide acoustic absorption.

Design solutions should seek to balance exposure of thermal mass with acceptable acoustics rather than allowing one consideration to dominate. A good compromise will take account of the likelihood of overheating, and a minimum acceptable acoustic standard, and will probably require some calculation and analysis.

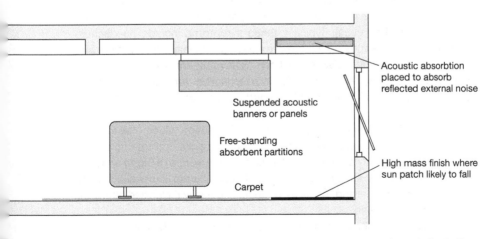

Figure 7.8 Acoustic absorbing banners and partitions providing absorption whilst leaving thermal mass in floor/ceiling slab exposed

Source: N. Baker

SERVICES AND CONTROLS – AIR CONDITIONING: AVOIDANCE OR REDUCTION

It is now well known that as a population, air-conditioned buildings use significantly more energy than naturally ventilated buildings. There is thus a large incentive to adopt as a strategy the avoidance of air conditioning. This will be facilitated by reducing gains as much as possible, and using the building to mitigate the effects by means of thermal mass and ventilation.

However, there may be some circumstances where air conditioning may be unavoidable. This demands two considerations – first, to minimise the air-conditioning load by passive means and, second, to provide the cooling and air handling by the most efficient plant available.

A third consideration is that air conditioning may only be needed in certain parts of the building and at certain times of the year. The way in which the transition is made is important – for example for a space for which natural ventilation is adequate for all but extreme hot weather – it is very important that the natural ventilation is discontinued when mechanical cooling is present. The strategy of partial air conditioning is often referred to as 'hybrid' or 'mixed mode'.

The design and specification of the services during retrofit will have great influence on the success of the building both for its energy performance and the provision of comfort. Here we concentrate on a few key design parameters that are likely to affect the comfort and well-being of the occupants. These mainly involve the interfaces between the occupant and the services, i.e. the heat (and 'coolth') emitters and the controls.

Sick Building Syndrome

Sick Building Syndrome (SBS) refers to the observation that in certain buildings, complaints from occupants of a particular set of symptoms are more common than usual. Typically these symptoms include headaches, irritation of the nose and throat, sore eyes and occasionally skin irritation. The type of building in which SBS is most commonly present is the deep, open-plan sealed building, relying heavily on mechanical ventilation (or full air conditioning), artificial lighting and with little or no personal environmental control.

It is also common in older buildings of this type which may indicate that it is due to a degradation of the mechanical services. In particular, air quality can be seriously compromised by poor maintenance of filters and poor cleanliness of ducts and grills. The use of recirculation can result in the distribution of bacterial and fungal pathogens, especially if clogged filters are automatically bypassed.

SBS is far less prevalent in shallow-plan naturally ventilated and daylit buildings. If the building to be retrofitted allows these passive features to be restored, it would be the best strategy. If the building to be retrofit is a deep-plan highly serviced building with a history of SBS, this presents a bigger challenge, and it would seriously suggest that at least the mechanical services need major maintenance or replacement and upgrading. The provision of user controls (discussed below) would also be essential.

Heat emitters

It has already been mentioned that improvements to the envelope insulation and ventilation control will not only lower the overall heat demand, but will also reduce the temperature differences within the external envelope. The absence of areas of high heat loss, and the lower demand overall, means that heat emitters such as low-temperature underfloor or radiant ceilings are viable. These have a high level of occupant satisfaction reported when used at low outputs. Both systems are also suitable for cooling, if it is considered that acceptable summer conditions cannot always be met by passive means.

The building may already have a perimeter fan-convector system that was designed to offset the originally high perimeter losses. If these losses have been reduced significantly due to fabric and glazing upgrades, their output can be reduced simply by reducing water temperature difference (ΔT). This improves boiler efficiency and is compatible with low-energy systems such as heat pumps and solar thermal.

Again due to the reduced perimeter losses, it may be better to remove the fan convectors and replace with simple panel radiators of lower output. These carry the significant comfort advantage that they are silent and do not contribute to dust distribution, and the operational advantage of low cost and low maintenance.

However, if fresh air is introduced by passive vents associated with the windows, these will remain a point of concentrated heat load, which now becomes a larger proportion of the total load. This must be dealt with in order to prevent local comfort problems, either by combining a controlled fresh air intake with the perimeter convectors or panel radiators, or by a stand-alone mechanical fresh air supply for use during low ambient temperatures. This can be demand controlled (by CO_2 concentration control point), and preheated, possibly with heat recovery from extracts in central zones. Heat recovery from exhausted ventilation air, together with very high levels of insulation, is an essential part of the Passive House approach for new buildings. This leads to very low auxiliary heating demand. A less extreme specification has been developed for retrofit called EnerFit.

Whilst cool floors are a viable option for small cooling loads where there are no carpets, they may present condensation risk if carpets are present, due to the temperature drop over the thickness of the carpet. Chilled beams and ceiling panels are a more familiar coolth emitters, but mechanical cooling should only be resorted to when passive means cannot achieve the minimum standards, and then only in 'hybrid' mode where mechanical cooling is applied intermittently and locally in response to peak conditions.

The comfort band can be significantly extended by the use of room fans to create air movement; typically a reduction of effective temperature of around 3°K can be achieved at practical air velocities. These can be ceiling mounted or located at desk level, the latter being preferable in that, like task lighting, they offer more personal control and thus another important adaptive opportunity. With ceiling fans used for cooling, there is also the danger that they will direct warmer stratified (and possibly more polluted) air down onto the occupant.

Artificial lighting (and integration with daylight)

Great progress has been made in luminous efficacy fluorescent light sources and, together with improved luminaire design, has led to a steady lowering of the installed power (W/m^2). The use of high frequency control gear has eliminated flicker and improvements in colour rendering have both contributed to improvement in visual comfort. This means that unless the building has received a recent lighting upgrade, the artificial lighting will be a prime candidate for the retrofit.

Task lighting

There is now widespread agreement that task lighting is potentially beneficial both for energy efficiency and visual comfort, but many lighting schemes fail to reach their full potential. This is because task lighting is often installed that provides local *lighting in addition* to high levels of general lighting. From an energy perspective, the purpose of task lighting is to allow a lowering of the ambient room illuminance (from say 300 lux to 150 lux), by providing higher illuminances at the workstation. Task lighting will also allow much greater use of daylight, since it will extend the time (and area of the floor) for which daylight provides sufficient light for circulation and non-demanding visual tasks.

If the task lighting is too powerful, and has no means of dimming control, this principle will be undermined, since high levels of workstation illuminance (say 750–1,000 lux, quite easily achievable with task lighting), due to adaption of the eye, may cause a background illuminance of 150 lux to appear inadequate. It goes without saying that the task light sources should be high efficacy fluorescent or LED; if incandescent sources such as tungsten halogen (still popular in domestic equipment) are used much of the energy saving will be lost.

Task lighting is also beneficial for visual comfort as, due to its controllability, it complies well with the principle of adaptive opportunity. It is particularly appropriate in workplaces where much of the information is communicated by luminous screens, where high levels of task illuminance are not needed or even undesirable.

Spatial design of lighting

Less progress has been made in the spatial design of lighting. Luminaire designers have concentrated on low-brightness sources,[3] often set in a suspended ceiling in a downlighter configuration. This results in a low-luminance ceiling and the perception of a hidden light source, with a rather nocturnal quality. Whilst this may be suitable for retail display, or a hotel foyer, it is not a good daytime working environment. This is because it does not emulate the luminance distribution of the natural sky condition and it is difficult to integrate with daylight.

There is now serious consideration being given to how visual cues, such as lighting colour and configuration, may be effective in influencing our natural circadian rhythms and hence health and well-being. Whilst this is a relatively new area, it seems likely that lighting schemes with high ceiling luminance are better for daytime occupants. This allows far easier integration with daylight, since diffuse sources with both upward and downward light output can supplement daylight. It is also compatible with occupant-controllable task lighting.

Controls

One of the principles of adaptive comfort is that people are more satisfied with their environment if they feel they have some measure of control over it. Efforts to engineer the perfect neutral environment have always failed to eliminate a residual level of dissatisfaction. There is a view (Baker and Standeven, 1997) that we have an innate desire to respond to our environment – the more neutral it becomes, the more sensitised we become to a particular parameter. The adaptive comfort principle, on the other hand, is to provide a 'good enough' environment, and provide the means by which the occupant can respond.

However, the principle of personal control over the environment has to be set against two constraints. First, in many situations occupants share the space with others, sometimes, in the case of open-plan offices, too many people for a consensus view to be found. In the case of cellular offices, occupied by two to four people, consensus is easier and personal control is more successful. This suggests that the open-plan solution is not ideal for the compromise between energy efficiency and health and well-being. However, the problem can be mitigated by providing some personal control – such as task lighting and desk fans, and permitting and even encouraging a more mobile working pattern, where occupants may use temporary locations away from their home station. This may involve shared facilities (warm-desking) and need not necessarily lead to a high space provision per occupant.

Second, personal control should not be allowed to seriously compromise energy efficiency. For example, opening a window because a heated room is too hot may achieve comfort, but will result in a waste of energy. Ironically, the invention of the thermostat has created this problem, as illustrated in Figure 7.9. Here, control technology can provide an answer; it is not difficult to integrate a window sensor with the local heating and ensure that the heating is discontinued, or at least set at a low level, as soon as the window is opened. This must not be seen as a 'punishment'; it is simply achieving the wishes of the occupant in a more energy-efficient way.

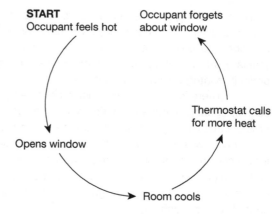

Figure 7.9 The impact of the thermostatic control to break the natural feedback loop in occupant control

Source: N. Baker

Another area of 'innocent' misuse is in lighting control. Whilst we are aware of too little illumination, we are much less aware of over-illumination – at least at levels found in buildings. Thus, it is not uncommon to find lights still switched on in a room when daylight has become completely adequate. The simple answer is a photo-sensitive control which fades the lighting gradually, but allows it to be overridden if for any reason an occupant feels the need for extra light. After some time delay, or coinciding with a particular disruption to the daily pattern such as lunch break, the lights will again be returned to off.

Caretaker controls

We have called this the 'polite caretaker' function. The control system antici-pates human interaction, tolerates a small measure of 'illogical' behaviour, but gently returns the building to a base level environmental status. Fortunately, modern technology, including IT and wireless control communication, has made sophisticated control of buildings possible and affordable. The development of digital control now means that individual actuators can be addressed with the minimum of infrastructure.

However, in the past, this has been used as an excuse to have a high level of central control and automation – the so-called 'smart' building. We see it rather differently. This wonderful facility should be used to welcome the adaptive behaviour of the occupant, but to ensure that it neither conflicts unreasonably with other occupants, nor the energy-efficient opera-tion of the building. Perhaps we should be moving towards the 'polite' building: polite to the occupant and polite to the environment.

THE INTEGRATED APPROACH

An underlying theme in this chapter has been that measures to improve energy efficiency can also improve comfort and well-being. Looked at another way, it follows that for a given level of comfort and satisfaction, certain retrofit specifications can lead to greater energy savings than others. The avoidance of air conditioning is a good example.

Another theme has been the provision of adaptive opportunity. Less energy will be used by providing a 'good enough' environment where a modest level of behavioural response is possible than attempting to provide a closely controlled 'ideal' environment.

Finally, it is clear that many measures have a mutually supportive effect – thus reconfiguring windows may improve daylighting, improve view, reduce glare and overheating, and open up possibilities for natural ventilation with intuitive occupant control. These multiple benefits may be more than just technically supporting, but also psychologically, i.e. the whole is greater than the sum of the parts. This is borne out by the findings of the 'forgiveness factor' by Building Use Studies (Leaman and Bordass, 2007). On the other hand, some actions may have undesirable effects on other environmental parameters – for example, exposing thermal mass can have adverse acoustic effects.

A further psychological factor is the perception of 'natural causes', which helps the occupant to understand the non-neutral environment and, we find, become more tolerant of it. Closely related to this is the degree to which the building allows contact with nature.

This suggests that the designer must not become too focused on any one environmental issue, but be aware of their interactions, and both the positive and negative consequences on the indoor environment as a whole.

NOTES

1   Defined as the ratio of the overall satisfaction score to the average score of individual environmental performance indicators.
2   It is predictable, using Sabine's formula: $T = 0.016\ V/\Sigma\alpha A$ where V is the volume of the room, α is the absorption coefficient of the room surface element and A is the area.
3   This refers to the apparent brightness of the luminaire when viewed at an oblique angle from a normal sitting position.

REFERENCES

Baker, N. and Standeven, M. (1997) 'A behavioral approach to thermal comfort assessment', *International Journal of Solar Energy*, 19: 21–35.
Fanger, P. O. (1970) *Thermal Comfort Analysis and Application in Environmental Engineering*, Vanloese: Danish Technical Press.
Guedes, M. C. and Matias, L. (2009) 'The use of adaptive thermal comfort criteria in office buildings: A study in Lisbon', *Renewable Energy*, 9(9): 2357–61.
Humphreys, M. (1978) 'Outdoor temperatures and comfort indoors', *Building Research Practice*, 6(2): 92.

Leaman, A. and Bordass, B. (2007) 'Are users more tolerant of "green" buildings?', *Building Research and Information*, 35(6): 662–73.

Papairi, K. (2004) 'Daylight perception', in K. Steemers and M. A. Steane (eds) *Environmental Diversity in Architecture*, London: Spon Press.

Rassia, S. Th., Alexiou, A. and Baker, N. V. (2010) 'Impacts of office design characteristics on human energy expenditure: developing a "KINESIS" model', *Energy Systems*, 2(1): 33–44.

8

ENERGY-EFFICIENT PRINCIPLES AND TECHNOLOGIES FOR RETROFITTING

Simon Burton

The most important part of environmental sustainability in a refurbishment project is minimising the use of energy, particularly energy from non-renewable sources, to reduce the contribution to climate change. There are two aspects: embodied energy in the refurbishment process and energy in use in the subsequent lifetime of the building. Reducing lifetime energy use is usually by far the larger component and the area in which the largest impact can be made in refurbishment, even reducing it to zero in some cases with the judicious use of renewable energy sources. This chapter summarises the actions that can be taken in the design to minimise energy use in the operation of a building being refurbished, in its fabric, services, controls, etc. Chapter 12 describes the other main areas where environmental sustainability can be improved, reducing water use, providing recycling facilities, choosing materials for the works and landscaping on and around the building.

Design for energy efficiency in a new building needs to start as early in the project as possible, at the briefing and concept stages, so that efficiency can influence the location, layout and basic design concepts and lead to the optimal solution. With a refurbishment project, the design needs to start with the existing building, what level of refurbishment is being considered, the energy efficiency of all elements or of those elements which will be retained and, where available and relevant, historical data on comfort and energy use in the building. The principles of energy efficiency and the technologies available will be virtually identical to those for new build, the challenge for the refurbishment designer is to select those technologies that maximise the use of what already exists, exploit the potential of the building and integrate new technologies that complement these and make the building as energy efficient as possible.

Refurbishment can treat all aspects of energy use, as the case studies in Chapter 15 show. The techniques and technologies used depend on the situation in the building, whether it is architecturally listed or has protected features, the condition, the use to which it will be put, the budget, the whims of the designers and many other factors. Designers need to consider the effects on energy use of all aspects of a refurbishment and consider how energy use can be minimised, bearing in mind the requirements of comfort for building users and likely building management capabilities.

The latter aspect of building management is highlighted in Chapter 14 which emphasises the importance of follow through from design intentions, through construction, to actual building operation. The designer must always consider how the building will be operated and ensure the design is appropriate for its likely future use.

ENERGY USE IN BUILDINGS

Energy is commonly used in buildings for:

- lighting, to supplement daylighting;
- heating, both space and water;
- active cooling, where necessary;
- mechanical ventilation, where natural ventilation is not possible;
- equipment, mostly IT-related, plus facilities such as lifts and catering;
- specialist equipment for production in factories;
- travel associated with the building (see Figure 8.1).

But renewable energy sources can also be retrofitted in buildings to reduce use of fossil fuel for:

- generating electricity;
- providing heat;
- supporting cooling.

These different aspects are not of course independent; many relate to each other and thus an integrated approach to design considering all related aspects will produce by far the best results.

For more detailed information and technical solutions, the *Handbook of Sustainable Refurbishment: Non-Domestic Buildings* (Baker, 2009) gives full descriptions of systems and technologies appropriate for energy-efficient refurbishment.

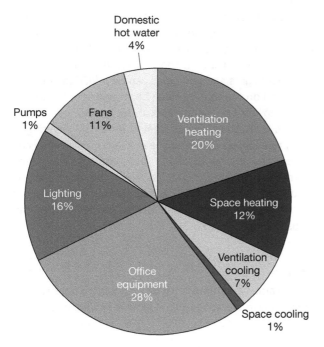

Figure 8.1 Breakdown of building energy use

Source: Simon Burton

The following sections give a general outline of typical approaches to optimising energy use in the different areas, which are available to designers involved in retrofitting and renovation. The case studies in Chapter 15 demonstrate the practical application of these approaches in different situations, why they were chosen and how they were incorporated.

LIGHTING

Electricity for lighting is one of the largest of the energy uses in offices and is very dependent on control systems and user operation.

Daylighting

The starting point in refurbishment design is to optimise the use of day-lighting by adequate window openings in the facade (see Chapter 9) and location of internal spaces to maximise occupant use of the daylighting. Use of daylighting from facades is closely associated with appropriate shading which is frequently necessary to reduce solar gain and to avoid glare for areas close to the windows. Poorly designed or poorly controlled shading may cause internal blinds to be used to stop sunlight but lead to the use of artificial lighting (the blinds down/lights on syndrome). Perforated and/or moveable external devices can be retrofitted to optimise these sometimes conflicting requirements. Carefully designed 'light shelves' can be retrofitted to window openings to reflect daylight further into a room while reducing intensity adjacent to the window. Roof lights and sun pipes can be added to corridors and internal rooms if located on the top floor and borrowed light brought in from adjacent spaces using glazed partitions, with high-level windows if privacy is required (Figure 8.2). Atria can be created either by covering in existing courtyards or by opening up a new space in a large block. Atria can allow daylight into the centres of large blocks to illuminate adjacent rooms and to provide an intermediate space for circulation and functions such as cafes and seating. In addition, atria effectively add insulation to the adjacent building facades and can also be used as part of the ventilation system, typically to expel exhaust air (see Figure 8.3).

Artificial lighting and control

The control system for the artificial lighting must be designed to optimise the use of daylighting and minimise electricity use. Lighting circuits need to be

Figure 8.2 Borrowed light in corridors and internal rooms

Source: Simon Burton

Figure 8.3 A new atrium attached to an existing building. Photovoltaic panels are used as shading

Source: Simon Burton

designed or modified to enable local zone control of reasonably small areas, including areas with good daylighting and separate rooms. Local photo controls to turn down or off lighting when daylight is adequate are widely used. Occupancy detection to turn off lighting when rooms become unoccupied and on and off for corridors can save a lot of lighting energy. Display lighting in retail buildings is a large energy user and heat generator and refurbishing with low-energy luminaires and control, where possible, will provide energy savings.

Low levels of background lighting with local desk task lights are an essential component of an efficient office lighting strategy. Naturally, low-energy luminaires in glare-free fittings should replace older inefficient systems. The use of LEDs (Light Emitting Diodes) is becoming very common as they are very efficient and have a long life. The control of security and emergency lighting, and lighting for cleaning, can be areas forgotten but significant energy users.

Lighting and particularly daylighting have to be considered in conjunction with heat loss in winter from larger window areas and excessive heat gain in summer depending on orientation, and the different requirements need to be balanced during the refurbishment design process.

HEATING

Most non-domestic buildings in temperate climates need heating for some periods of the day, at least in winter, while internal and solar heat gains will provide much of the required heating demand in refurbished buildings, as long as they have been insulated and draught-proofed.

Solar and other heat gain

There will almost always be some contribution from sun entering the building, passive solar gain, depending on the orientation, building form and over-shading, but often little can be done in refurbishment to increase this. Larger windows on south and east elevations can be effective, depending on overshadowing from adjacent buildings. However, exposure to sun is more likely to cause overheating in office buildings (see below) though shading systems that allow morning easterly sun to enter the building but restrict sun on the south and western elevations can work well. Internal heat gains from occupants and equipment will always provide a large part of the heating demand, even if new equipment is chosen to minimise electricity use and thus the heat emitted. These heat gains need to be managed to make best use of them by controlling ventilation and, with mechanical ventilation, installing heat recovery systems. Installing the most efficient equipment is always very important, as although the heat gain is useful for space heating, it is not the most efficient way of providing it, and overheating in summer is a common problem. Computer suites and some manufacturing equipment will generate large amounts of surplus heat which can be used for space heating other parts of a building. Careful design is necessary to assess the available heat, the timing and the efficiency of heat exchange and transfer systems to achieve successful use.

Insulation

Adding insulation of the external elements of the building is the best way to reduce heating demand, and all floors, walls, roofs and windows should be insulated where possible. All the elements of new facades will normally be well insulated. Retrofitted facades can have insulation added in the cavity, if one exists, externally or internally to the solid parts, and windows can be replaced by new ones. External wall insulation and cavity fill are normally preferred as this leaves the thermal mass of the wall on the inside, useful for temperature control. However, external insulation may not be possible if the facade is protected or listed or for other reasons, and internal insulation must then be used. In all cases, analysis of condensation risk should be carried out so that interstitial condensation and moisture build up are avoided. Vapour barriers should always be placed on the warm side of insulation and the installation specified carefully to seal joints and maintain a complete barrier. Insulation can be added to get the building up to Passive House standard and this will result in heating only being required for warm-up in the early morning on cold winter days. New glazing systems use double- or triple-glazing, with argon fill and low-emissivity (low-e) coating, in thermally broken frames, and these are highly effective in reducing heat loss and cold radiation to occupants near windows. Secondary glazing can easily be added where windows or glazing are not to be replaced and can be a good solution for reducing both heat loss and noise from adjacent sources. Draught-proofing of renovated facades needs careful attention. Roofing in courtyards or previously open spaces to make atria provides an element of insulation to adjacent walls and windows.

Treatment of cold-bridges is likely to be important to reduce heat loss and avoid localised condensation, unless complete renovation back to the building structure is proposed. Complete over-cladding will eliminate cold-bridges but internal insulation can leave cold-bridges at floors and vertical members, which will need careful treatment. Replacing windows can leave a serious cold-bridge at the edges and mullions, which will need at least some insulation added. Balconies and other protruding structural elements can also be very difficult to treat.

Insulation of top floors is a standard process for ventilated pitched roofs or flat roofs where there is good ventilation below the waterproof roofing. If the waterproofing is not being replaced, waterproof insulation placed on top of the waterproof layer is a very useful and effective way of providing insulation, as well as protecting the existing waterproofing. It needs to be weighted down with slabs or stone, and reduction in parapet or safety rail heights must be considered. Where new insulation is placed below a new waterproof layer, great care must be taken in placing a vapour barrier and encouraging ventilation to avoid interstitial condensation. Placing insulation directly below the waterproof membrane will run the risk of degradation of the waterproofing due to large temperature changes and thermal gradients as the roof structure will not have a moderating effect.

Insulation of ground floors may be possible depending on the degree of refurbishment and should follow normal new build practice.

Heating systems

Replacement of old direct electric heating systems is always a high primary energy-saving change, as will be replacing old non-condensing gas and oil boilers. Boiler efficiencies can be increased from as low as 50 per cent to 80 per cent and above. Heat may be available from local district heating networks and buildings can be retrofitted to connect to these. This is likely to be a good and energy-efficient method of providing heating, and even cooling via absorption chillers. Biomass boilers can provide all or part of building heat demand and can be retrofitted if fuel delivery and storage space can be found. Traditional oil or gas boilers are the normal way to provide heat, via the ventilation system, underfloor or traditional radiators, convectors and fan coils. The highest efficiency is obtained by using condensing boilers with distribution systems that allow low temperatures to provide sufficient heat, such as underfloor or with large heat emitters. Modern heat pumps can provide efficient heating systems if carefully designed, ideally using underfloor heating layouts to improve efficiency, and well controlled. Ground-sourced heat pumps using boreholes or extensive buried pipe systems can be retrofitted if there is appropriate space and ground conditions around the building and these can provide a really efficient system (see Figure 8.4). Heat recovery in ventilation systems is widely used, due to the large amount of heat arising from occupants and office and other installed equipment and heat exchanger efficiencies can be 80 per cent and above. There are likely to be opportunities to insulate existing pipework and heat storage vessels where heating systems are not being replaced, to reduce heat losses and improve efficiency.

Combined heat and power systems can be retrofitted to commercial buildings to give efficient operation in buildings with sufficient heat demand. Generated electricity, if surplus to requirements in the building, can be exported to the grid but the system will only be efficient if much of the heat is needed in the building. One option is to use the heat to drive absorption chillers but this is unlikely to be efficient overall compared with conventional chillers under normal circumstances.

Figure 8.4 Ground-sourced heating or cooling

Source: Simon Burton

Controls

Efficient use of heating systems requires installation of appropriate controls so that heat is only supplied where and when it is needed and certainly to avoid conflict with cooling systems. Time controls with optimal start and thermostatic controls are essential and easily retrofitted. Zoning of different areas to enable use of solar and other heat gains, and local control are recommended, where adjustments are being made or new heating systems installed. Where windows are openable, micro controls that cut off local heating (and cooling supply) when windows are opened can be used. Automated control by Building Management Systems (BMS) is normally used in commercial buildings and if correctly designed, installed, commissioned and operated provides for efficient use. Unfortunately, this is often not the case and the design must take into account the likely use and whether the management personnel will be able to understand the system and operate it adequately. Chapter 7 describes the importance of occupants' ability to change local conditions being a significant factor in achieving adaptive comfort and any control system should ideally include the facility for local occupant control.

Domestic hot water

Domestic hot water systems can be very wasteful of energy if hot water is constantly circulated round buildings to supply different areas, due to heat losses and pump power. Uninsulated pipework can also cause overheating in summer, leading to more cooling demand. Localised hot water generators using gas or electricity with short pipe runs should be retrofitted in most circumstances to supply individual or grouped facilities.

COOLING

Non-domestic buildings usually need to expel heat generated by internal activities and solar gain in summer, and climate change is likely to make this a more important factor in the future for building designers.

Minimising internal heat gain

Any cooling strategy should start by minimising the internal heat gains from equipment, mostly IT, by buying efficient equipment and ensuring it is turned off when not in use. Specific heat generators such as IT suites and print rooms should be located separately so that they can be separately vented or cooled. Solar gains need to be controlled by external or internal shading which can be retrofitted to existing facades as well as incorporated into new ones and this is an important energy-saving and comfort-enhancing technology (see Figure 8.5). There are many options for external shading including

Figure 8.5 Solar shading fixed to an existing facade
Source: Simon Burton

fixed overhangs and louvres, moveable louvres, retractable blinds and vertical fins and shading for east and west facades. Shading not only reduces unwanted solar gain but also avoids glare, particularly necessary with the wide use of computer screens. Internal blinds may have to be used on architecturally protected facades where changes in appearance are not allowed, but these are not as effective at reducing solar gain as external ones. Flat roofs that are exposed to the sun should be treated with reflective surfaces both to reduce heat gain through the roof but also to reduce the degradation of the waterproof covering.

Natural cooling

Natural ventilation using wind and stack effects can be adequate to maintain reasonable temperatures in small and carefully designed and refurbished buildings in cooler climates, and mechanical extraction can be used to boost air movements if necessary. Passive cooling strategies can be enhanced by designing in exposed thermal mass to be cooled by colder night air and absorb excess heat during the day, for example, by removing suspended ceilings to expose existing concrete slabs. Night cooling strategies require careful design of secure openings throughout the building to enable cold night air to pass through and reduce the temperatures of the thermal mass. Mechanical ventilation will make this more effective, though requiring the use of electricity. Control is necessary to operate a night cooling strategy for only when and for how long it is useful, otherwise overcooling can make the building too cold first thing in the morning. Night cooling can be increased with the addition of phase-change material (PCM) in ceilings and walls, which comes in various forms, and which increases the amount of heat stored in the day and removed at night (see Figure 8.6). Ceiling-mounted fans are more often used in warmer climates for increasing air movement and hence the cooling effect, but can be used in temperate climates. Retrofitting fans is very dependent on ceiling heights but they can enable higher temperatures of up to 2°C to be felt as comfortable, in the absence of other air movement.

Active cooling

Mechanical ventilation with extract and/or supply can provide adequate removal of heat gains for much of the year but many large commercial buildings will require active cooling to maintain adequate temperatures in hot weather. Low-energy active cooling systems can use ground sourcing where the ground supplies cooler water to cool the ventilation air directly or as a cold source for an electric chiller integrated into a full mechanical system. Chillers will otherwise normally need to have rooftop cooling towers, wet systems giving the highest efficiency but requiring more careful maintenance to avoid health risks. Another low-energy cooling source can be absorption chillers using heat from solar collectors, district heating systems and waste heat from other sources.

Figure 8.6 Phase-change material ready to install

Source: Simon Burton

Displacement ventilation systems with chilled beams or chilled ceiling panels may be able to be retrofitted depending on the space available and are considered to be a low-energy system compared with others, due to the large emitting area and consequent acceptance of higher coolant temperatures. Condensation risk analysis is necessary to avoid condensation on the cold beams under extreme conditions.

Controls are once again central to energy-efficient cooling to avoid unnecessarily low temperatures and to ensure that active cooling does not operate when free cooling is possible nor at the same time as heating. Cooling set temperatures should be designed to be readily adjustable to allow them to be raised when ambient external temperatures allow occupants to be comfortable with higher temperatures (see Chapter 7).

VENTILATION

Natural ventilation

Narrow-plan buildings with the ability for cross ventilation may be able to be comfortable with natural ventilation and this will provide the best energy efficiency where usable. Natural ventilation needs careful design to enable wind and stack-driven air movement to provide enough fresh air for a healthy internal environment and to remove excess heat gain in summer. Draught-free controllable openings in the facades can be windows or low- or high-level vents and these latter can be wind pressure controlled to control the amount of wind-driven ventilation. Mechanical extract only can be installed to provide minimum air movement when there is not enough wind.

Mechanical ventilation

Mechanical ventilation with the supply of tempered air and extract of stale air is frequently necessary and a requirement in commercial buildings to give adequate indoor air quality and comfort. Retrofitting of ductwork may be restrained by existing floor to ceiling heights but removal of suspended ceilings can provide more space, as well as exposing the thermal mass. Variable air volume (VAV) systems with low-energy fans and pumps replacing older systems will provide energy savings. Another solution is to draw fresh air from controllable facade openings through the building using ceiling voids and corridors with extract fans usually on the roof. Heating and cooling will then need to be supplied by traditional radiators, convectors or underfloor heating, and chilled beams. Low-energy fans retrofitted to replace older models in existing ventilation systems will save significant energy use. Older duct systems may require high fan power to overcome restrictions and old filters, and replacement can greatly reduce the amount of fan power. Well-designed control systems are vitally important for both comfort and efficiency, with CO_2 sensors to control flow rates frequently used to ensure good air quality with minimum air flows.

Mixed-mode ventilation, where natural ventilation can give adequate conditions for much of the year with full mechanical ventilation and air conditioning turned on only when necessary can, if properly designed and operated, be a low-energy solution. Careful design of the natural ventilation system is necessary so that it does not impede efficient operation of the mechanical system. It is the changeover period from air conditioning to natural ventilation that needs to be carefully organised to avoid apparent discomfort to occupants.

ELECTRICITY-USING EQUIPMENT

Retrofitting should be the opportunity to install the most efficient equipment which will both reduce energy consumption and the related cooling demand, particularly important in summer. Mechanical services equipment, fans and pumps, are available which use a fraction of the energy of older models and can frequently replace older components without difficulty. Office equipment, servers, computers and printers are major energy users in offices and should be chosen for maximum energy efficiency and programmed to ensure that they are turned off when not in use. The need for well-controlled efficient lighting is covered above.

Replacement lifts and escalators can all be chosen to be low-energy versions. Kitchen equipment for food storage, cooking, serving and washing up can all be chosen, and systems designed, to minimise energy use. Other equipment is used in retail facilities and service and manufacturing buildings, from washing and drying machines, ATMs, tills and printers, to manufacturing equipment. These tend to be specialist areas but all worthy of study to minimise energy use.

TRANSPORT, CYCLE FACILITIES

There is a growing interest in the energy used in travel to work, and the options for reducing this range from home working and the related aspect of 'hot-desking', to location next to good public transport, reducing car parking availability and encouraging employees to cycle to work. These also reduce congestion on the road network and reduce atmospheric pollution. The building designer can encourage energy-efficient transport patterns by including sheltered cycle racks, showers, and other facilities in their buildings. Frequently the number of car parking spaces previously available in the pre-refurbishment building have been reduced to discourage car commuting, which also makes space available for other activities such as cycle facilities and waste storage for recycling.

ENERGY METERING AND MONITORING

Whilst design for energy efficiency in the various energy-using elements of the building is vital, their operation and use must be understood and controlled. The designer should therefore build in separate meters for energy-using equipment and different areas of a building to enable energy use to be monitored and excessive use analysed and hopefully reduced. Electricity meters can be retrofitted to existing circuits depending on the layout and gas meters depending again on layout and on appropriate pipe runs to avoid turbulence. They should be linked to the BMS displays to enable the facilities manager to understand and minimise energy use in their day-to-day work. If typical energy use figures are established for different functions and zones, local meters will identify equipment or controls failure or wasteful human intervention, which can be investigated.

RENEWABLES

Some renewable energy sources are very applicable to existing buildings and can often be a cost-effective retrofit measure, more so than some energy-saving measures. Retrofitting of biomass boilers, photovoltaic arrays and solar water heating are becoming common features of renovated commercial buildings. Biomass boilers, often backed up by gas boilers, can be retrofitted or built into new heating systems and provide the majority of heating from this renewable resource. Biomass boiler capacity is often designed to meet only the base heat load of the building with top-up using the gas back-up boiler as necessary, due to the slower response time and lower flexibility of biomass boilers compared with gas boilers. Adequate fuel supply arrangements, in-building storage and boiler supply and ash disposal need to be designed into projects for biomass systems to be used successfully. Photovoltaic arrays, normally roof mounted, depend on generally south-facing and unshaded space availability and visual issues on architecturally listed buildings and in conservation areas (see Figure 8.7). They have the advantage that all the electricity generated can normally be used in the building as it is likely to be occupied at peak generation times in the day. Surplus electricity should always be fed into the mains grid. Solar water heating using flat plate or evacuated tube collectors can be retrofitted on roofs depending again on space and architectural limitations. Domestic hot water again will be generated by the solar panels mostly when it is in demand in the building.

Ground-sourced heating and cooling systems, daylighting and passive solar gain, as described in the sections above, are also considered as renewable sources and these can make a large contribution to energy saving in a building. There has been some interest in retrofitting small wind turbines to buildings to generate electricity, but few, if any, good examples exist. Installing larger turbines adjacent to buildings on industrial estates and

Figure 8.7 Large-scale photovoltaic modules mounted on a roof
Source: © AS Solar GmbH

in parking areas is likely to be a more successful option for tapping wind energy to generate power for a building.

REFERENCE

Baker, N. (2009) *Handbook of Sustainable Refurbishment: Non-Domestic Buildings*, London: Earthscan.

9
THE IMPORTANCE OF FACADE DESIGN
David Richards

Refurbishment, especially in the commercial market, can be described as the realignment of a building's durability with long-term economic value. That is to say – how do you reinvigorate a building so it will last longer and be worth more. Sustainability sits neatly at the apex of both these issues. The act of refurbishment is inherently sustainable in that the waste of demolished material and embodied carbon of a new structure are avoided and presumably a once 'tired' building is once again active and useful.

The facade is an essential element of the building's performance and of its inherent value. Aesthetically the facade gives a building image, responding to and creating context, and giving the lasting impression of value. In this way it defines the longevity of a building.

It is also the key environmental filter of a project – heat, light, air, sound and views all pass through the facade. Not only does the building envelope impact the performance of the building in embodied and operational carbon terms but it also influences the performance of the people working in the building. Their comfort and concentration are fundamentally related to the facade.

This chapter examines the key facade typologies involved in refurbishment and critiques the technical issues surrounding the design and repair of cladding on existing buildings with a focus on cold and temperate climates.

TYPOLOGIES

Refurbishment of the building envelope is always specific to each project. Refurbishment often involves a mixture of approaches depending on the different facade types involved.

For simplicity, envelope refurbishment can be considered in four generic typologies:

- over-cladding – installing a new envelope directly over the existing;
- re-cladding – removing the existing facade and installing a new system;
- refurbishment – repairing and reglazing an existing facade;
- retained facades – keeping and repairing an existing facade whilst replacing the building behind it.

Each has its own drivers, risks and rewards, which are examined below.

Over-cladding

Over-cladding, as the name suggests, involves leaving the existing cladding system in place and installing a new system over the top of it. Whilst the building will retain its geometry and more than likely its mix of solidity and transparency, an over-clad building can look very different in colour and texture offering the chance to reinvent its image (see Figure 9.1).

There may be a number of reasons to choose over-cladding:

- It is usually the lowest cost option because the costs of demolition are avoided and the new system can take advantage of the performance of the existing facade system.
- Internal work should be at a minimum as the new facade can often be installed with the building's tenants left in place. This keeps rental income flowing while the work is done.
- Savings in discarded demolition materials should make it a low embodied carbon option.
- The building's overall look remains similar so it may offer a smooth ride through the local planning process.

However, over-cladding is not as simple as it sounds. Most buildings that are over-clad have an existing facade which is old and often predates many of the advances in performance and complexity of modern facades. There are a number of issues that should be considered on a project-by-project basis:

- Survey the existing facade – the condition and performance of the existing facade need to be well understood because they will inform the design of the over-cladding system. This means they must be surveyed with careful attention paid to their fixing details, insulation and glazing details. Having made this examination it is worth reconfirming that over-cladding is still the right solution!
- Make sure the structure is strong enough – in effect, over-cladding doubles (or more) the structural load of the facade. It is critical to analyse the existing structure to make sure it is strong enough or can be reinforced.
- To meet modern energy codes and best practice it is likely that the glazing will need to be replaced. This means examination of

the interfaces between the existing facade, the new facade and the new glazed elements.

- The introduction of a new facade will change the vapour pressures in the facade build-up which means condensation risk will occur in a different place. This must be analysed by calculation.
- The fixing system and brackets of the new system will need to attach either to or through the existing facade. This means they become weak spots both structurally and thermally. These also need to be detailed and analysed.
- The approach to maintenance of the facade will need to be checked. The existing arrangements may no longer be appropriate after the over-cladding. For example, cleaning cradles may no longer reach far enough.

In short, over-cladding offers a low-cost and effective way of upgrading the image and performance of a building envelope. But be aware that success or failure is all in the details so the decision to over-clad should not be taken too quickly or without thinking through the detail.

Re-cladding

Re-cladding is perhaps the most common option in refurbishment. The existing building envelope is completely removed and replaced by a new system. Within the geometric constraints of the existing building, re-cladding can reinvent a building's image and value. It also gives the opportunity to redefine its environmental performance.

Figure 9.1 Guy's Hospital, London. The new facade was installed in front of the original one

Source: Arup

Guy's Hospital, London, was designed by Watkins Gray and built in 1979 and it was the tallest hospital in the world until 2008. It is made out of two towers, which were originally clad with precast concrete. Arup and Peynore and Prasad Architects carried out a design for the over-cladding of the existing concrete facade in order to improve the thermal performance, repair the concrete panels and replace the windows that had reached the end of their service life. The new facade is installed in front of the original one, so that the building could be kept operating with only minimal disruptions. The design had to be fully coordinated with the findings of surveys, in terms of room for installation and structural capacity to support the additional loads.

It is worth noting that this example from the medical world was chosen because there are very few successful over-clad schemes in the commercial office sector.

Re-cladding may be chosen for a number of reasons:

- It allows a complete rebranding and repositioning of a building's value in its market.
- The use of daylight and thermal performance of the building can be redefined.
- A completely new facade is likely to be a manufactured product so can be of higher quality with minimum material wastage.
- The improvements in thermal performance can open up opportunities to use more sustainable and efficient forms of air conditioning.
- A new facade with excellent thermal performance gives the chance to consider natural ventilation and other low-energy mechanical systems.
- Replacing the building envelope offers the opportunity to add area and hence value to the building and move the envelope line out or up.

Replacing the facade on an existing structure is in many ways the simplest of the options because it comes closest to starting from scratch. Nevertheless, it is not without its challenges:

- The edge conditions of the structure need to be surveyed to allow the support for the new facade to be detailed.
- The structure of the building needs to be evaluated to ensure it can support the weight of a new facade.
- The floor-to-floor heights of the facade are retained and may be a constraint. This may limit the use of daylight and natural ventilation.

- A new facade is likely to have the highest embodied carbon because the components, at present, are all made from new material. However, this should be balanced against the improved operational performance and the fact that the embodied carbon of a new building structure has been saved.
- It is likely that a new facade will cost more than other options, although this needs to be balanced against better cost certainty and lower technical risk compared to other options.
- The building will almost certainly need the tenants moved out, although there are exceptions for particularly modular facades where the system can be replaced progressively out of hours.

Re-cladding offers the best opportunity to radically alter a building's image and value and make the most comprehensive improvements in the building's performance. But the benefits come at cost – both in terms of capital cost and lost rental income.

Window upgrade

Changing the windows is probably the simplest option available. Whilst it can have significant impact on carbon performance it has the least impact on external visual appearance. This may be a good thing if a building has heritage value or it may be driven by cost.

Window upgrade may be chosen for a number of reasons:

- It has the lowest cost because the solid elements of the facade are left intact.
- The cultural value of the building, if any, is retained.
- It is likely that planning permission will not be needed or will be minimal, although this can vary locally.
- The work can probably be done with the building's tenants left in place either completely or with a rolling decant. This may help entice a tenant to stay for relatively modest investment.
- The thermal performance of the windows can be improved by the use of double- or triple-glazing and coatings and by improving the air permeability of the glazing system.

Whilst window replacement is relatively simple it should be done with care and the challenges should be recognised:

- Window replacement tends to have more impact on the internal environment than it does on external appearance. This needs to be evaluated in terms of the impact on brand and value.
- If the building is of heritage value the visual appearance of the glazing may be constrained by planning or conservation legislation. This may include narrow framing profiles, glass colour and reflectance, etc.

- The interfaces with the remainder of the facade need to be considered and detailed carefully to ensure they perform thermally as well as possible.
- Only replacing the glazing means there is no opportunity to improve the performance of the rest of the facade – poor insulation values, thermal bridges, etc. will remain.
- It is likely that daylight levels will be reduced because new, more insulating glazing will use coatings and have more layers, thereby reducing the transmission of light (and heat). This needs to be reviewed in the round as an overall operational carbon issue.
- Only replacing the windows will almost certainly need the use of site-applied wet sealants which needs supervision to ensure high quality.

Window replacement offers a low-cost alternative with relatively modest impacts on carbon performance. Under certain planning or conservation constraints it may be the only viable option (see Figure 9.2).

Figure 9.2
Lloyds of London. The floor-to-ceiling glazing was replaced in order to improve the thermal/solar performance and to allow for more daylight

Source: Arup

Lloyd's of London, designed by the then Richard Rogers & Partners and built in 1986, received Grade I listing in 2011 – the youngest structure ever to obtain this status. Between 2011 and 2012 the floor-to-ceiling glazing was replaced in order to improve the thermal/solar performance and to allow for more daylight. The original glazing units were made out of three panels of 'sparkle glass' and one panel of vision glass (image at the bottom right of Figure 9.2). This ratio was modified to 50 per cent also to improve the feeling of transparency, without losing the language of the facade (top right). The new units – produced by recycling the existing material – were installed from the inside due to the configuration of the building, with no interruptions of the business inside the building (bottom left).

Retained facades

Some buildings have such huge heritage value in their facades that they need to be preserved because it is their appearance and cultural history that gives the buildings their value. This often goes hand-in-hand with planning and conservation restrictions (see Figure 9.3).

Facade retention offers many advantages and opportunities:

- Heritage buildings often establish a specific and positive brand. For example the historic stone buildings offer a feeling of permanence and trust that is of huge value.
- Retaining facades preserves often prominent buildings with public value to society and conserves craftsmanship which simply cannot be recreated today.
- Providing the scope of refurbishment is wide, there is opportunity to improve carbon performance by replacing windows and adding insulation.
- Preserving a historic facade and satisfying planning and conservation concerns may gain planning 'credit' which can be used to allow a more expansive approach to the remainder of the building, possibly, for example, additional area and value.

At the same time, working on heritage facades is a skilled and somewhat specialised practice and there are a number of issues to consider:

- Overall carbon performance improvements may be limited. Glazing system selections are often limited visually – narrow, thermally less efficient framing, limited colour tints to glazing.
- If the floor slabs behind the facade are also retained thermal bridging at slab edges is likely to remain.

Figure 9.3 Unilever House, London. Facade retention, the stone-clad, steel-framed facade was upgraded with insulation and vapour barriers

Source: © Arup; interior © Mike O'Dwyer (from Arup Asset Bank)

> **Unilever House, London.** Retaining the facade of Unilever House in London (1932, Grade II listed) made a crucial contribution to a Royal Institute of British Architects (RIBA) award-winning, Building Research Establishment Environmental Assessment Method (BREEAM) 'Excellent' project. The stone-clad, steel-framed facade was upgraded with insulation and vapour barriers to comply with Building Regulations for thermal performance, and original windows were replaced with high-specification glazing.

- Adding insulation to opaque elements will change the vapour pressures in the facade build-up and this needs to be examined by calculation to ensure that condensation does not occur in places which increase the risk of water or frost damage.
- Facade retention always necessitates a high degree of on-site working and fewer manufactured components. This needs to be supervised and attention paid to quality.

It is worth mentioning in passing an extension of facade retention – dismantling and rebuilding historic facades. This is driven either by a need to replace materials or to change dimensions to suit a new use (see Figure 9.4). Whilst sharing many of the attributes of retained facades it further diminishes the embodied carbon benefits of retention because energy is expended in the work needed and inevitably new material is needed in the reconstruction.

Integrating historic facades is an exciting art form. Its sustainability strengths are associated with the preservation of a building social value and less about carbon performance. The technical challenges should not be underestimated.

COMMON ISSUES

Throughout the various typologies – and the grey areas in between them – there are common issues to be dealt with. Indeed, many of these issues are common to the facades on newly constructed buildings too. Four of the major issues are discussed below – natural ventilation, modern insulation levels, detailing and whole carbon.

Natural ventilation
One of the great things about designing for temperate climates is that you can open the windows. It makes for a better experience for occupants and it saves energy because the air conditioning can be turned off. And natural ventilation is likely to become more prevalent as our automotive stock shifts from petrol-

Figure 9.4

0–211 Piccadilly, London. The Portland stone and brick facade was dismantled and rebuilt 1.5 metres higher, on top of a new ground and first floor structure

Source: © Arup (Thomas Pearson)

210–211 Piccadilly, London. The Portland stone and brick facade of 210–211 Piccadilly (1891) was dismantled and rebuilt 1.5 metres higher, on top of a new ground and first floor structure. This allowed for continuous floor plates between the corner plot and adjacent new build office buildings on all sides.

driven cars to hybrid and electric vehicles. Our streets will become quieter and cleaner and it will be a much more pleasant experience to open the windows. Successful natural ventilation is based on two principles:

1   That comfort should not just be measured by air temperature (i.e. the temperature measured by a thermometer held in the air) but instead by Operative Temperature which takes account of the radiant temperature of surrounding surfaces.
2   That people will tolerate warmer temperatures inside as the outside temperature rises – so called adaptive comfort – if they can control their own environment and have a connection to outside (see Chapter 7).

This approach has a number of important consequences for the design of facades:

* limiting solar gain in summer;
* ensuring ventilation in winter;
* controlling surface temperatures;
* designing mechanisms for opening the facade.

Solar gain
Natural ventilation, as the name suggests, uses air naturally available outside to meet both the cooling and ventilation needs of a building. The defining issue for the facade of an office building is usually associated with cooling, even in cooler temperate climates. In comparison with air conditioning the temperature of the air outside is much warmer so its cooling capacity is lower. This means that heat gains need to be controlled to make sure that it does not get too hot inside. All the heat gains need to be tuned to make this work – lighting, heat from computers and people – need to be considered. The overall peak needs to be in the order of 60 W/m^2 if exposed thermal mass can be employed with night-time ventilation to recharge it. Of the normal likely heat gains the largest and most variable is solar gain through the facade. This needs analysis for the specific site and climate but it is likely that the overall solar transmission will need to be limited to below 15 per cent.

This raises the spectre of fully glazed buildings, which could be the subject of an entire book in itself, but is less applicable to refurbishment than it is to new build. Suffice to say, the solar transmission needed is unlikely to be achievable with full-height glazing and at best will be very expensive. Consideration also needs to be given to the issue of glare.

The result of the analysis of solar transmission is likely to produce glazed areas of 35–50 per cent, even in a temperate climate. This can vary from the lower end on the dominant sunny facade (south in the northern hemisphere, north in the southern) to the high end on the shady facade (north in the northern, south in the southern). These figures can be increased if the project's budget will run to external shading and even more if the shading can be moveable.

If a project's budget allows for it, external shading can provide an elegant and effective way of allowing optimum daylight, clear views and good control of solar gain. Physically it can run the gamut between simple fixed overhangs to moveable louvre systems or deployable external roller blinds.

Incidentally, in a temperate climate heating in this analysis is also important. It is not usually the issue which defines the glazing ratio but once that ratio is established for heat gain the level of insulation needs to be examined to minimise energy use and carbon footprint.

Winter ventilation

Similarly in winter there needs to be a mechanism available to ventilate enough fresh air to avoid the space feeling stuffy. This is a much lower quantity than needed for cooling in the summer and can be supplied in a number of ways:

- Operable vents with a minimum winter position – this approach is likely to work well in large transient spaces – atria, entrance lobbies, etc. – where draughts are less of a concern. However, in occupied offices the draughts from operable windows can be significant.
- Trickle vents – very small openings, so called because they allow a small 'trickle' of air through driven by external wind pressure. These can work very well for relatively limited floor depths – say 4 or 5 metres from facade to the limit of the space. Beyond this a trickle vent will struggle to supply enough air without beginning to become a source of draughts. For this reason, trickle vents tend to be more popular in residential than commercial construction because the depths of space are narrower.
- A mechanical ventilation system – where plan depths go beyond 4 or 5 metres a mechanical system is likely to be needed to ensure ventilation air is evenly distributed. A mechanical system will also offer the advantage of high-efficiency heat recovery. It also means the facade can be focused on being well sealed when the windows are closed, i.e. just having two modes of operation – open or closed – rather than a third intermediate mode.

Surface temperatures

The temperature of the inside surface of facades needs to be understood to make sure it is not causing discomfort.

Controlling surface temperatures can be considered in two parts – opaque elements and glazed areas. Opaque elements simply need to have sufficient insulation to ensure that their surface temperature is close enough to room temperature to avoid discomfort. In temperate climates which are particularly sunny, care should be taken to use the right external temperature to calculate the internal surface temperature – the external surface temperature will be raised by the sun's radiation. In extremes it may be worth considering a separated external layer with a ventilated cavity to remove the solar component of the heat gain using a similar set of details to rain screen cladding.

Insulation levels are discussed below in further detail.

Glazed areas are where much higher surface temperatures can occur – as high as 45°C – so much more care needs to be taken. A combination of the following elements should be considered – the same elements you might use for a new building:

- double- or triple-glazing;
- low-emissivity (low-e) coatings to minimise heat gain and loss;
- internal blinds to limit solar gain and reduce the internal surface temperature (in this case the blind itself when it is down);
- a ventilated cavity either on the inside or better still on the outside.

The options need to be assessed by computational analysis to estimate surface temperatures and resulting comfort.

Mechanisms

A further consideration, if natural ventilation is to be considered, is designing operable elements to let air in and out. Consideration needs to be given to a number of issues:

- the use of high-level and low-level openings to encourage ventilation driven by natural buoyancy, or single tall openings;
- separation of a motorised high-level night ventilation opening (to recharge thermal mass) from a lower vent possibly manually controlled, for daytime use;
- the use of motorised drivers and automatic actuators to open windows via the building controls system;
- depth and width of framing to incorporate independent framing for the operable elements and thermal break.

Insulation and detailing

At the nexus of sustainability, temperate climates and commercial buildings lies the issue of insulation levels. How much insulation is needed? The standards for insulation vary from place to place depending on the local energy codes and their level of sophistication. In general, codes are moving towards having greater levels of insulation to minimise heating loads.

It is worth noting, and being slightly wary of, the argument that for a commercial building you can have less insulation because the building is full of heat (people and computers) and you need to let the heat out to reduce cooling loads. This is partially true but it should be remembered that computer loads in particular are likely to drop as cloud computing becomes more popular and processing power and screen technology become more energy efficient. There is also an argument that buildings need the flexibility to change use to other sectors such as residential where insulation levels need to be higher.

The insulation levels required need to be determined case by case using an energy model of the building in accordance with the local energy code. Note that this determines a minimum legal position (i.e. any less insulation would be illegal!). To go beyond this will need further analysis and a determination of goals and available budget.

It is possible to improve the thermal performance of uninsulated existing facades by adding insulation either externally, internally, or within the wall build-up (see Figure 9.5). Depending on the thickness of the insulation, different U-values can be achieved. The right balance between performance, cost and room required for the insulation needs to be found on a project-by-project basis.

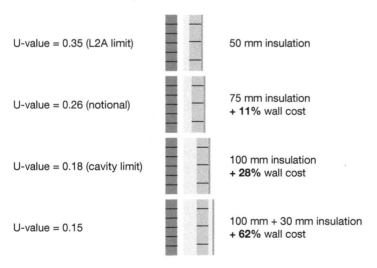

U-value = 0.35 (L2A limit) — 50 mm insulation

U-value = 0.26 (notional) — 75 mm insulation
+ **11%** wall cost

U-value = 0.18 (cavity limit) — 100 mm insulation
+ **28%** wall cost

U-value = 0.15 — 100 mm + 30 mm insulation
+ **62%** wall cost

Figure 9.5 Improving the thermal performance of uninsulated existing facades by adding insulation

Source: Arup

153

A particular concern with new and refurbished facades is framing. With increasingly high-performance glass available the framing becomes a weak point for heat loss, heat gain and discomfort. For new windows or curtain walling this can be limited by the use of thermal breaks – insulating plastic inserts to break the metal on the outside from that on the inside. For retained historic facades where the framing must be kept for architectural heritage the impact must be accounted for in the heating and cooling calculations for the building. In such situations it may be possible to introduce secondary glazing to go some way to improving the thermal performance, although this brings with it its own set of issues.

The thermal performance of frames can be assessed only by means of finite elements thermal analyses. Figure 9.6 shows a two-dimensional analysis of a typical intermediate transom, separating the vision glazing area from the insulated spandrel zone – in this case a shadow box. The image on the left shows the geometry of the model and the thermal conductivity of the different materials assumed. The output of the analysis is the temperature distribution along the detail – image on the right – and the heat flow through it. The frame is responsible for the highest rate of heat losses, and consequently there is a significant variation in the temperature gradient in that area.

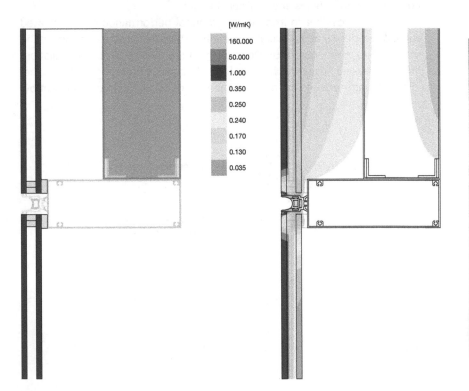

Figure 9.6 Thermal analysis of framing details

Source: Arup

A further issue which merits close analysis is thermal bridging. A common area for this is where the facade meets or passes the end of the floor slab. In many older buildings there is a direct thermal bridge at this junction which is a cause of heat loss and condensation risk on the inside surfaces. If the building is being re-clad or over-clad it may be possible to run the new insulated cladding past the slab edge. However, if the facade is being kept then the only option is to insulate the slab itself on the inside. This can be hard to achieve architecturally if the slab is exposed or floor-to-floor heights are restricted.

A note of caution should also be given with masonry walls. Insulated masonry construction by its nature needs ties to pass through the insulation layer to tie the inner and outer layers of the wall together. Each of these ties is a thermal bridge and there can be a lot of them. With highly insulated walls the ties begin to dominate the performance of the overall wall to the point where adding more insulation has diminishing returns.

Moisture movement is an important factor in any wall construction but when adding insulation to a facade extra care needs to be taken. By adding insulation on the inside – usually the only place to do so practically – the vapour pressures and temperature in the original facade elements are changed. This can lead to new mechanisms for decay such as freeze/thaw cycles in masonry or rot in timber with now moist fabric cut off from the warmth of the building.

EMERGING ISSUES

Sustainability as a driver for commercial refurbishment is still evolving and the level of sophistication varies by geography from being a market differentiator to being a legal commitment. The facade industry is also an industry which is rapidly evolving with much of the development being driven by sustainability – for example, the increasing complexity of framing systems is a response to increasingly stringent thermal performance criteria.

Looking across the spectrum of the two – an increasingly sophisticated view of sustainability and constant improvements in facade technologies – there are a number of emerging themes that are worth mentioning and which may become important for the near- and far-term future.

Energy performance standards
In the more developed markets where sustainability has been an active component of design and refurbishment for a number of years, there is an emerging focus on real-time performance – i.e. actual energy use not just calculated predictions.

One consequence of this for the building envelope has been an increased emphasis on actual performance. For example, for some time this has meant that the solar performance of glazing must be tested and certified to industry standard codes. Also sustainability rating schemes like BREEAM are now requiring as-built testing for air leakage and post-completion thermal image scanning to prove insulation levels and thermal bridging standards are met.

The airtightness of existing building can be measured by means of air pressure tests in which the whole building is pressurised by means of fans installed at the entrance of the building (see Figure 9.7). The locations where the main air leaks occur can be identified by spraying smoke with a smoke-gun: where the air barrier is not continuous there will be concentration of smoke flowing out.

Figure 9.7 Airtightness testing
Source: Arup

To check the thermal performance of an existing facade, we can use thermal imaging, which provides a clear idea of the temperature gradient along the building envelope (see Figure 9.8). By taking a 'simple picture', it is possible to identify the location of thermal bridges, corresponding to the areas where high temperature gradients occur. As the finite element analysis shows, the critical areas are at frames' locations.

For the refurbishment market we are beginning to see legislation emerging that will set minimum energy performance standards for existing buildings as well as new construction. As part of energy legislation the UK government will soon set a minimum performance standard below which a building will not be legally rentable. This will radically change the economics of refurbishment, making it mandatory to improve the energy performance of a building prior to re-renting. The building envelope is a central element of this shift because it plays such a fundamental role in defining energy performance.

In cities where the commercial office market is well established there is a further challenge for the facade industry. The economics of most buildings is driven by minimising so called 'voids' – the periods of time when

Figure 9.8 Checking the thermal performance of an existing facade using thermal imaging

Source: Arup

there are no tenants and therefore no income. Refurbishment can create voids because the building, or large chunks of it, will be out of action. It may also be the catalyst for a tenant to activate clauses which allow them to break their lease and relocate to newer higher performing buildings. So the challenge to the facade industry is how to respond to this with innovative ways of refurbishing or replacing the facade of existing buildings without displacing the tenant and indeed giving them compelling reasons to stay for the long term.

The glazing of the Empire State Building in New York City did not provide the thermal and solar performance required for a modern building. The solution adopted to have a good balance between reusing the existing glass and improving its performance at the same time, was to dismantle the existing insulated units, and reassemble them with a high-performance film suspended in the cavity (see Figure 9.9). But it was necessary to keep the building running as normal, due to its enormous value. In order to do so, tailored logistics solutions were necessary: an on-site assembly line was installed at the fifth floor, so that it was very quick to open the glazed units, insert the film and reseal the units. More than 6,500 windows were replaced outside of working hours (nights and weekends) in approximately 12 months.

Embodied carbon

As the greenhouse gases emitted in running buildings reduce so the emissions associated with building or refurbishing them becomes more important.

This makes 'embodied carbon' an emerging theme in all areas of design and construction of our built environment. How much carbon is emitted from the supply chain and the process of building itself? The answer needs to consider the energy used and carbon emitted in extracting raw materials, in turning them into products, transporting them to construction sites, in the process of installing them and in the frequency of their

Figure 9.9
The Empire State
Building, New York.
More than 6,500
windows were
replaced outside of
working hours

Source: Arup

replacement cycles. Emissions from end of life demolition and waste treatment can also be important.

The building envelope occupies a large proportion of the embodied carbon of a building because it uses high-energy materials such as glass, steel and aluminium all of which need high temperatures, and hence energy, to process and which are bulky to transport. In new buildings with unitised facade systems this could be as much as 30 per cent of the overall embodied carbon. In a similar refurbished building, where the structure is retained along with its original carbon footprint, the building envelope can be responsible for as much as 50 per cent although this is highly dependent upon the detailed nature of the refurbishment of a project's embodied carbon at the time of construction. Additionally, facades, unlike structure, have a replacement cycle of twenty to thirty years, which in overall terms makes their lifetime impact even higher.

So embodied carbon is an important issue for the building envelope and the next few years will see much activity to try and tackle the traditionally high values facade systems produce. These activities are likely to include:

- Developing framing materials with lower embodied carbon such as low-temperature plastics and timber.
- Lowering the processing energy associated with glass.
- Increasing the recycling of glass from facade systems.
- Developing and applying concrete precast systems that make better use of cement alternatives.
- Reusing steel and aluminium profiles from existing facades. There are issues to satisfy regarding warranting of reused sections which the industry will need to address (i.e. design to enable materials to be recycled efficiently or even reused. Closing the material loop will reduce whole-life embodied carbon of facade systems.)
- Extending the life of double- or triple-glazing by developing alternate longer-life cavity closure systems. It is worth noting that the failure of glazing seals is often the thing that triggers the replacement of facade systems.

Beyond embodied carbon, the theme explored in this chapter will grow to 'embodied environmental impacts' more widely. Aspects such as embodied water, toxicity, summer smog creation, biodiversity, etc. can all be measured when considering the impacts 'embodied' in the materials and associated supply chain of the facades industry. As assessment methods become standardised and information more widely available it will be our responsibility as practitioners to address these impacts as part of good design practice.

CONCLUDING COMMENTS

The building envelope has a vital role to play in the sustainability of any commercial building. It is the medium through which people experience the outside environment, it controls raw energy flows into and out of the building and substantial carbon is emitted in its production.

It is also one of the key elements that defines a building's value which for refurbished buildings can take many different forms – heritage, reinvigoration or complete renewal.

This chapter has outlined the key typologies and the issues which need to be considered in developing efficient and elegant facades – one of the first steps towards the sustainable buildings our planet needs for the future.

10
ENVIRONMENTAL ASSESSMENT RATING SCHEMES

Simon Burton

Ever since environment and sustainability were recognised as important issues in the construction and operation of buildings, the industry has been swamped with claims and counterclaims and 'green wash' has been common and has hampered respect for serious progress in demonstrating how buildings can make a more sustainable contribution to environmental issues. Part of the problem has been the complexity of defining the relative importance of different environmental issues, measuring the environmental impact of components and proving that the product is actually what it is claimed to be. Much of the problem stems from the difficulty and costs of going seriously green so that some players have attempted to promote their environmental status by over-exaggerating the positive aspects of their developments. The first environmental rating schemes in the building sector were set up as frameworks for assessment and comparison, and started defining what was sustainable and how to measure it. This was intended to give an objective assessment of as many environmental issues as possible related to building construction and use.

Since early developments some twenty-five years ago when the first Building Research Establishment Assessment Scheme (BREEAM) was launched in the UK for new buildings, there has been a proliferation of schemes in different countries, for different building types, for different building works (new builds, fit-outs, refurbishments), as well as for the operation of buildings. Although schemes have historically been used mostly to assess new build schemes, the use and importance of specific existing building and refurbishment schemes is becoming more significant.

There is no doubt that the development of assessment schemes has been an important driver for raising consciousness about environmental issues in buildings and improving the sustainability of building construction and use. Assessment schemes also fulfil a very important requirement in modern society in helping to set standards and targets and demonstrate achievement in practice.

HOW ENVIRONMENTAL ASSESSMENT SCHEMES ARE USED

Building environmental assessment schemes are used by almost all of the major stakeholders in the building industry, from planning authorities, private and public sector developers and occupiers, through to designers and on-site constructors. The results of assessments may also have a life post-construction in promoting and marketing a building, in helping to benchmark buildings against each other and throughout their useful life. Some schemes are specifically designed to measure the sustainability of a building's use and can be used as a management tool throughout the life of the building.

Planning and legislation

Local, regional and national planning authorities can use assessment schemes to demand higher environmental standards for refurbishment and new building projects in their jurisdiction. National laws and regulations are in place in all countries to address a range of environmental issues, but local authorities may have good reasons for wanting developments in their areas to achieve higher environmental standards. Whilst the legality of this will vary by country, developers may be encouraged to adopt high standards to make gaining planning permission more straightforward. Some local authorities set specific targets to be achieved to obtain planning permission.

Public and private clients

It is common for clients of all sorts, both building developers and occupiers, to include a requirement for achieving certain environmental standards in their policies and briefs. The public sector is often required to be in the vanguard on certain issues to promote government policies, best practice, or kick-start industry in a particular direction. Thus, including the requirement for a certain environmental rating to be achieved in all new and refurbished government buildings, be they owned or rented, directly or indirectly funded, can satisfy a number of these criteria. Owner-occupiers are likely to have a direct interest in specifying high environmental standards for their premises, both for their corporate image and for staff satisfaction and retention. Developers, even if driven primarily by cost and value, normally now recognise that those renting or buying their building may well be looking for minimum environmental standards. Whilst a high environmental rating is likely to be well down the list of priorities for most tenants, it does provide a differentiator from other similar buildings in the same area.

Designers

Thus, many design briefs now include either a requirement to achieve a certain environmental standard, or at least a general requirement to produce an environmentally friendly building. For the designers, an environmental assessment scheme will provide a checklist of issues, an explanation of what the compliance requirements are for each, different options for achieving a set standard and normally an overall rating which allows the building to be benchmarked against others. As with many building issues, environmental issues are best considered from the start of a project rather than at the end of the design process, when design changes to achieve a higher environmental standard will be costly or not achievable. The design team can thus use the scheme as a part of the design development process, and involve all the participants in the decision-making on environmental issues. The use of schemes and checklists within design teams including all disciplines is thought to have been a significant contributor to raising environmental awareness across the building industry.

Construction and commissioning

As we know, environmental design does not stop with drawings and specifications: it must be carried through the construction, commissioning and building use phases. This is recognised by most environmental assessment schemes which require the completed building to be assessed rather than its design only. Most environmental assessment schemes also directly address the specific impacts associated with the actual construction of the building, and the significance of adequate commissioning. While most of the assessment components will be built into the construction contract, the assessment results are also useful to the on-site staff to check compliance and make sure the best environmental solutions are used.

Marketing and publicity

Environmental ratings are also widely used by those marketing buildings to potential clients. For example, BREEAM ratings are commonly given in marketing material and even on billboards. Today, a high environmental rating is synonymous with a responsible organisation, be it a private business, a government or any other types of enterprise such as major events, World Cups, Olympics, etc.

TYPES OF ASSESSMENT SCHEMES

Different schemes have been found to be necessary for different building types (housing, offices, retail, etc.), activities (new build, fit-out, refurbishment, in use), and regions (temperate, tropical, arid, etc.) to focus on the most relevant issues and to reduce the complexity of the scheme as far as possible. Earlier versions of BREEAM used to include multiple schemes, up to nine different schemes for non-domestic building types as well as schemes for housing. Some schemes have a more general approach (see later). Different countries have different climates, resources, traditions, as well as their own standards and regulations. Since assessment schemes aim to recognise achievements above the regulatory minimum, different countries have either developed their own assessment methods or have adapted those from other countries to suit their locality.

Schemes produce different outputs, from checklists to single or multiple ratings. They are administered in different ways, the most robust and trustworthy schemes arguably being those where claims are verified by independent third parties, typically accredited professionals and/or the scheme administrator. Numerous schemes also require specially trained and accredited individuals to assess projects.

ISSUES COVERED IN SCHEMES

Environmental aspects and impacts from buildings cover a range of areas from global, to local and social issues. Typically the following issues are included in assessment schemes:

- energy uses and CO_2 emissions, for all building uses;
- renewable energy sources in or on the building, in its vicinity or used by the building;
- embodied energy, including materials and construction works;
- environmental impact of materials used and their durability;
- health and well-being of building occupants, including commissioning of heating, ventilation and air-conditioning (HVAC) systems, access to daylight, volatile organic chemicals (VOCs) emissions, etc.;
- water use by all the facilities associated with the building, including landscaping;
- waste disposal from activities associated with the building;
- transport impacts related to accessing the building;
- land, water and atmospheric pollution from building-related emissions, including light pollution, heat island, storm-water runoff, etc.;
- site selection, local ecological damages and opportunities for improvement;
- management of the building's construction impacts, including erosion control, emissions from construction plant, impacts on neighbours;
- management of facilities when the building is in use, including pest management, supply chain, energy and water metering;

- the use of innovative technologies or solutions is also recognised by some schemes.

For some issues setting a standard is easy, e.g. all timber to be sourced from a certified sustainable source, whilst for other materials an environmental life cycle analysis is mandated and typically takes into account energy used in extraction and manufacture, scarcity of the basic resource, pollution in extraction and manufacture, use of recycled material, transport to site, longevity of the product, energy in use, recyclability at the end of its life and other factors. Information collection is a lengthy process and the results can vary greatly between different manufacturers. Thus, many assessment schemes rely on reference documents and other best practice standards which collect data and assess common products.

Some issues can be assessed simply on a yes/no basis, e.g. the provision of cycle facilities, the use of brownfield sites, and many others can be rated against country regulations and best practice standards, e.g. energy use, health and well-being, or water use. Of course, specific measurement and rating standards also have to be set down for most issues.

Weighting, and in most instances aggregating, different environmental aspects and impacts is always a difficult area for assessment schemes, in particular if the objective is to produce a single rating at the end of the assessment (e.g. BREEAM and Leadership in Energy and Environmental Design or LEED), rather than a spreadsheet output giving the assessment results for each individual issue or category, e.g. the French HQE system. There is obviously no objective way to compare or aggregate CO_2 emissions with local ecological improvements, so different schemes have come up with different approaches using experts or opinion surveys to derive a 'consensus' approach.

Some schemes include mandatory requirements which must be satisfied to achieve certification. For example, LEED includes seven Minimum Program Requirements (MPRs) which must be satisfied for a project to be eligible for certification. In addition, each LEED environmental category also includes 'prerequisite' requirements which must be met but are not rewarded with any points. BREEAM includes different Minimum Standards depending on the level of certification sought.

DIFFERENT AVAILABLE SCHEMES

BREEAM is generally thought to be the first major building environmental assessment scheme and has led to the development of assessment schemes around the world. Many of these can be used for the assessment of refurbishment of commercial buildings and some are briefly described below.

BREEAM

The Building Research Establishment Environmental Assessment Method (BREEAM) is used widely in the UK and across the world in schemes developed for a particular country, but also in the form of BREEAM International. BREEAM also allows the bespoke assessment of unusual buildings, where a unique, bespoke assessment scheme is developed for an individual building. BREEAM requires a formal assessment to be carried out by a qualified assessor, and verified by the scheme's administrator, the UK's Building Research Establishment (BRE). A global rating is determined by awarding scores for achievements at the post-construction stage against a series of 'credit' requirements. Each credit addresses a particular environmental issue and is grouped with other credits under several environmental categories (e.g. health and well-being, energy, etc.) Each category is weighted according to an importance assigned by BRE to that category, which is then converted to a unique rating from 'Pass' to 'Outstanding'.

BREEAM is used in a range of formats from country-specific schemes, adapted for local conditions, to international schemes intended for the certification of individual projects anywhere in the world.

International refurbishment projects can be assessed as follows:

- major refurbishment projects only, the BREEAM International New Construction 2013 scheme;
- all refurbishment and fit-out projects, the BREEAM Europe Commercial 2009 or BREEAM International Bespoke 2010 schemes.

Traditionally BREEAM has included refurbishment, fit-out and new build assessments as part of the same unique scheme as was the case with BREEAM 2008. BRE is currently developing a new stand-alone scheme for the assessment of non-domestic building refurbishment titled 'BREEAM Non-Domestic Refurbishment 2014'. This new version of BREEAM will provide a dedicated scheme for non-domestic refurbishment and fit-out, and will run alongside other BREEAM schemes such BREEAM New Construction and BREEAM In-Use.

LEED

Leadership in Energy and Environmental Design (LEED) was originally developed for use in the USA but is now widely used across the world. The scheme is administered by the US Green Building Council (GBC) and the Green Building Certification Institute (GBCI). For major refurbishments, LEED New Construction (NC) and LEED Core and Shell (CS) may be used, depending on the area occupied by tenants and the building owner.

The LEED certification process involves five steps:

1   Choosing which rating system to use – see above for major refurbishments.

2   Formal registration with the administrator – when registration is completed, the assessment is accessible to the project team online via the 'LEED Online' service.

3   Uploading evidence demonstrating compliance with LEED requirements, and submitting the certification application online for review by the administrator.

4   The administrator's review process, which differs slightly for each project type.

5   Certification – the certification decision is sent, which can be either accepted or appealed against. An affirmative decision signifies that the building is LEED certified.

The aggregated number of points a project earns determines the level of LEED certification the project will receive: Certified, Silver, Gold or Platinum.

Ska

Ska Rating is an environmental assessment tool that assesses the sustainability of the fit-outs of office premises. The scheme is operated by the UK's Royal Institute of Chartered Surveyors (RICS) and is thus relevant to some refurbishment projects. A version for retail fit-out is under development.

Project teams interested in fitting out spaces in an environmentally sustainable way can use the Ska Rating method to:

- carry out an informal self-assessment of the environmental performance of their fit-out;
- commission a quality-assured assessment and certificate from a RICS-accredited Ska assessor;
- obtain clear guidance on good practice in fit-out and how to implement it;
- benchmark the performance of fit-outs against each other and the rest of the industry.

One particularity of the scheme is that the first step of the assessment process consists of scoping the environmental issues of relevance to the project being assessed, based on the works to be undertaken. Unlike most other major assessment schemes, environmental issues which are not relevant to the works are dismissed and not considered by the assessment. On completion of the assessment, a rating of Bronze, Silver or Gold may be achieved.

Green Globes

The Green Globes system is a Canadian building environmental design and management tool. It delivers an online assessment protocol, rating system and guidance for green building design, operation and management through third-party verification. Of the available tools, the 'Design of New Buildings and Significant Renovation' can be used for refurbishment projects in the USA and Canada. Green Globes uses an interactive process that includes online software tools for the design team to provide evidence of compliance, with third-party assessments performed on site by seasoned personnel. The rating provided is in the form of one to four globes, indicating the increasing levels of environmental performance achieved.

CASBEE

The Comprehensive Assessment Scheme for Building Environmental Efficiency is a tool for assessing and rating the environmental performance of buildings and the built environment, developed by the Japan Sustainable Building Consortium and the Japan Green Building Council. There are a number of different schemes including some for refurbishment, though they were not available in English in 2013. Buildings are rated from 'Excellent' through to 'Poor'.

Green Star Australia

Green Star is Australia's mark of quality for the design and construction of sustainable buildings, fit-outs and communities. Green Star has grown into a comprehensive rating system for all types of projects and building sectors.

Green Star Certification is a formal process which involves a project using a Green Star rating tool to guide the design or construction process during which a documentation-based submission is collated as proof of this achievement. The Green Building Council of Australia commissions a panel of third-party Certified Assessors to validate that the documentation for all claimed credits is in adherence with the Compliance Requirements.

'Green Star – Office Design v2' rating tool evaluates the environmental potential of the design of commercial offices (base buildings), for both new and refurbished projects. Specific schemes are available for retail and industrial buildings.

Other environmental assessment schemes

Although BREEAM and LEED are by far the most widely used schemes in the world, there are many other environmental assessment schemes, typically developed for a particular country or language. They include the following, which is not an exhaustive list:

- HQE for 'High Environmental Quality' developed and mostly used in France;
- DGNB for 'German Sustainable Building Council' which originated in Germany and is also used in Central and Eastern Europe;

- Valideo a Belgian 'sustainable construction certification system' also used in Luxembourg;
- Minergie, the Swiss scheme, which focuses on occupants' comfort and energy consumption;
- the Chinese '3-star' scheme which is the Government's official evaluation standard for green buildings.

THE FUTURE OF ENVIRONMENTAL ASSESSMENT

The use of formal environmental assessment schemes is now common in many countries for both new commercial buildings and those being refurbished. The benefits of applying and using the outcomes of such assessments are relevant to most actors in the buildings sector; therefore, their use is steadily increasing and likely to continue to do so. We are likely to see more separate schemes developed for different aspects – climates, new build, major refurbishment, fit-out and in-use. We are also seeing a convergence in the approach adopted by the major schemes, making them no longer building use sector specific. Further changes are likely to be the inclusion of more environmental issues as the industry develops and innovations become best practice, the adaptation potential of a building to climate change as discussed in Chapter 3, and possible changes to weighting and rating systems in response to changing priorities.

11

ENERGY AND COMFORT MODELLING TOOLS

Ljubomir Jankovic

The process of sustainable retrofit needs to start by developing a model of the building before the retrofit. This initial model needs to be calibrated using energy bills or other information, such as data from instrumental monitoring, if available. A calibrated model is then used as an experimental tool to investigate different design decisions through simulation. We will explain individual steps of this process that the designer needs to take in order to develop a sustainable retrofit solution.

A book entitled *Designing Zero Carbon Buildings Using Dynamic Simulation Methods* (Jankovic, 2012a) explains that:

> modelling is defined as making a logic machine, which represents the material properties of the building and physics processes in it. Simulation is then defined as numerical experimentation with the model so as to investigate its response to changing conditions inside and outside the building.

In some retrofit projects it may not be possible or useful to base the modelling on the existing building, for instance, if it is derelict or if it will be stripped down to frame or if it will have a considerable change of use. In those cases, a model of a future building will be needed, which will be equivalent to creating a model of a new building.

MODELLING BEFORE RETROFIT

If the majority of the existing building were to be retained, rather than stripped down to frame, then an essential step in sustainable retrofit would be to establish a reference case against which the performance improvements can be measured. This can be achieved by creating a simulation model of the existing building, and subsequently by tuning the model performance to give results within a specified error in comparison with the actual building performance. This section explains how to obtain information on the actual performance of the building, and the next section explains how to tune the simulation model.

Obtaining details on the existing building

In the best-case scenario, all details on the existing building that are needed for the creation of the simulation model will be available: architectural CAD drawings; specifications of the building envelope components, such as walls, floors, roofs, glazing; specifications of the mechanical and electrical (M&E) systems, such as heating system type, heating, ventilation and air-conditioning (HVAC) plant technical specification including efficiency; specification and luminous efficacy of lighting; etc.

In the worst-case scenario none of these details will be available, and they will need to be obtained through a survey. Drawings will need to be generated on the basis of an accurate survey of the building geometry. Construction types of the building envelope components can be obtained on the basis of the year of construction and applicable Building Regulations of that time. This will not, however, necessarily give accurate information about the composition of the external wall, insulation thickness, etc. and can lead to inaccuracies in the simulation model.

More accurate information about the construction of the exterior walls can be obtained by using a borescope, an instrument that is used to inspect a construction through a small hole. The tip of the instrument is fitted with a light source and a refractive prism, which enables viewing perpendicular to its longitudinal axis. Therefore, by looking through it, we can see what is on the side of the instrument's tip. By inserting this instrument gradually into a pre-drilled hole and by recording its position and the material types observed through it, the user can determine the composition and thickness of individual layers, and thus obtain the specification of the construction type of the external wall.

Creating a simulation model

The creation of the simulation model involves the following steps:

1   entering the building geometry and setting the orientation;
2   setting the location and choosing the climate data file;
3   defining constructions and assigning them to individual components of the building envelope;
4   defining internal thermal conditions, internal heat gains, and infiltration and ventilation losses, and assigning these to individual zones/rooms;

5   setting time profiles of heating, cooling and internal gains;

6   entering the heating and cooling system specifications, such as system type, fuel type, seasonal efficiency, etc.

The above steps are necessary in order to define the simplest model that is ready to run. This kind of model can be enhanced by combining, for instance, models of air movement, daylighting and electrical lighting, inclusion of HVAC systems, etc.

Most simulation models can import computer-aided design (CAD) information, which can speed up the creation of geometry in the simulation model. As simulation models deal with 3D volumes rather than with 2D line drawings, the volumes need to be either imported in 3D, or regenerated from an imported 2D plan. Experience shows that importing 3D from CAD software can be problematic, as some surfaces can disappear in the process, resulting in a thermally open building envelope in the simulation model. Such a model will not run until the envelope is repaired to form a complete enclosure.

Obtaining energy performance data

As a minimum requirement, details of the actual performance of the building can be obtained from annual energy bills. However, this information will not give an insight into dynamic performance of the building as result of continuous external and internal influences.

Much more detailed performance data can be obtained from instrumental monitoring of the building, in which key parameters are measured and recorded in regular time intervals of between five and fifteen minutes. Longer recording intervals are likely to miss some of the information content related to the dynamics of building performance.

The parameters to be monitored would typically include external air temperature, solar radiation, internal air temperatures in representative spaces, sub-metered energy for heating, cooling, ventilation, lighting, plug loads, as well as occupancy patterns and others. If the building uses comfort cooling, then the relative humidity of the outside and inside air also needs to be monitored.

The choice of a data logger for monitoring would dictate the choice of sensors that are compatible with it. Modern data loggers are network-enabled and can be accessed remotely over the internet.

Air temperature and relative humidity sensors are quite common, and would be typically on the product list of the data logger supplier. Solar radiation meters, called pyranometers, need to be sourced from specialist suppliers and their compatibility with the data logger needs to be checked.

Sub-metering of electricity consumption can be done with clamp-on current transformers, which convert electromagnetic field around the electricity conductor into a voltage signal that is proportional to the current in the electrical circuit. Although these devices are non-invasive, they need to be mounted on a live phase, and therefore they require a split-core cable to be mounted on. Some clamp transducers come with split-core attachment kits that make the mounting easier. Some fixed electricity meters are monitored every half hour by the electricity supplier, and that information may be readily available on request without the need for additional monitoring.

Sub-metering of gas energy consumption can be carried out in several different ways. The easiest way is to source a gas meter with a pulsed output, with a specified pulse value in units of gas flow. Another way is to attach to the gas meter an external photoelectric sensor, which detects the movement of the gas meter needle. This is a non-invasive method but it may be difficult to set up.

Occupancy monitoring can be carried out using simple passive infrared (PIR) detectors. These sensors provide simple, normally-open or normally-closed, voltage-free outputs that are compatible with most data loggers. If installed in a representative place in the building, such as an aisle in an open-plan office, the number of pulses accumulated over the sampling interval will be proportional to the number of people and level of activity in the corresponding space. This information can then be used to generate activity profiles for insertion in the simulation model.

Baseline energy assessment

CIBSE TM22 is a structured method for energy assessment and reporting (CIBSE, 2006). The method can be used to carry out: overall energy performance analysis for a simple building; general building assessment for a multi-zone building with special uses; and evaluation of the system performance. In the context of sustainable retrofit, this method can be used for establishing baseline energy consumption of the building before the retrofit.

Having supplied information on metered energy and on the floor area to the TM22 spreadsheet, TM22 will generate an overall assessment showing how the actual building compares to energy consumption benchmarks for 'Good' and 'Typical' practice. This information is instrumental in early decision-making on the priorities and the strategy for retrofit. An example of TM22 assessment result is shown in Figure 11.1.

Information obtained from TM22 can also be used for improving the accuracy of the dynamic simulation model through a process of model calibration. TM22 can also be used for post-retrofit monitoring and analysis of energy and carbon performance, including the influence of unregulated loads.

Model calibration

The performance of a simulation model that has just been created can be considerably different from the performance of the actual building. This discrepancy between the simulated and actual performance is called a 'performance gap', and it is one of the major problems in building design.

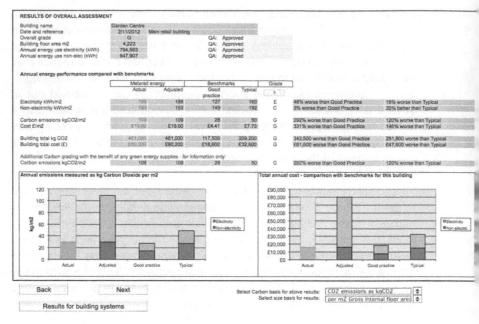

Figure 11.1 Baseline energy performance assessment with CIBSE TM22

Source: Data: Ljubomir Jankovic; Software: CIBSE (2006)

The performance gap can be considerably reduced in the case of retrofit, where the building already exists, and so do the records of its energy consumption. Depending on the performance information available, the objectives of calibration can be as follows:

1   to minimise the discrepancy between the simulated and actual annual energy consumption, based on annual energy bills;
2   to minimise the discrepancy between the simulated and actual temperature fluctuations in the building on an hourly basis, based on detailed instrumental monitoring.

In the former case, a relative error between the simulated and actual performance is calculated as shown in Equation 1, and the objective of calibration is to reduce error ε below a specified value, for instance, below 1 per cent.

$$\varepsilon = \frac{(simulated - actual)}{actual} \qquad (1)$$

In the latter case, a root mean squared error ε is calculated as shown in Equation 2, and the objective of calibration is to reduce the error below a certain temperature difference, for instance below 0.5°C, within a sample size of N data points.

$$\varepsilon = \sqrt{\frac{\sum_{1}^{N}(simulated - actual)^2}{N}} \qquad (2)$$

The reader can find an example of both annual and hourly calibration in a paper by Jankovic and Huws (2012).

The calibration needs to be carried out using a specially assembled or modified weather data file. There are different types of weather data files that are used with simulation models, such as Design Summer Year (DSY), Test Reference Year (TRY) and others. Details of these files are beyond the scope of this chapter; however, a useful overview of different types of weather files and their effects on simulation results was published by Crawley (1998).

In general, weather data files are aggregated using data from different calendar years. Whilst such aggregation is useful for exposing the simulation model to a range of different weather conditions that occurred over a number of years, all combined into a single year, this does not help with the model calibration. In the case of dynamic calibration with data obtained from monitoring, it is necessary to insert the external monitored parameters into the weather data file, thus making a bespoke file for the particular location and time period.

Weather data files with file extension 'EPW' (Energy Plus Weather) are particularly suitable for customisation, as they are in the plain text format and can be edited with a number of different text editors or spreadsheet programs. The EPW format description is available from the US Department of Energy (US DOE, 2013b). The simulation runs for calibration purposes will need to be carried out only for the time period that corresponds to the inserted data, which may be less than a year.

Calibration method

A calibration method that is used frequently is analogous to bracketing in artillery fire, in which the first shell is aimed beyond the target and the second shell is aimed short of the target, so as to establish the initial range (bracket) in which the target is. The process is then repeated, reducing the overshoot and undershoot by adjusting the angle of the weapon, until the target is hit.

Taking this analogy back into the context of simulation, the overshoot and undershoot are achieved by adjusting either the heating set temperature, or the infiltration rate, or some other suitable parameter in the simulation model. These parameters need to be adjusted one at a time, so as to establish one-to-one correspondence between the change of the parameter value and the resultant change of the simulation model output. The process is then repeated, making the overshoot and undershoot smaller and smaller, and thus reducing the bracket in which the target is. The calibration process is completed when the error between the simulated and the actual performance becomes smaller than the target error set out before the start of calibration.

DESCRIPTION OF THE DIFFERENT TOOLS AVAILABLE

There are several hundred building modelling tools (US DOE, 2013a), but how would we choose the ones that are appropriate for sustainable retrofit? Dynamic modelling tools perform detailed analysis of heat transfer, energy performance and thermal comfort, using transient heat transfer principles, and provide high-granularity information of the predicted building performance. Simulations are carried out in time steps of one hour or less over a full year, or over a chosen shorter period.

Steady-state evaluation tools provide low-granularity information and are often used for checking the compliance with standards. Calculations are typically carried out on a monthly average basis, using steady-state heat transfer.

This section gives an overview of different tool types. The choice of examples is not based on a comprehensive review of all available; different authors may prefer different tools, and therefore the reader is encouraged to investigate alternatives before selecting the tools for their own simulation design work.

Large companies also have their own in-house modelling programs, and individual world regions have developed models that are suitable for the respective climates.

More detailed descriptions of these software tools are beyond the scope of this chapter; however, several of these tools and their use are discussed in more detail in Jankovic (2012a).

Dynamic modelling tools

DesignBuilder

DesignBuilder (DBS, 2013) uses EnergyPlus (LBL-SRG, 2013), described in the next section, as its main simulation engine. The user interface is dashboard-like (Figure 11.2), where the majority of controls are available from a single screen. The dashboard contains tools for visualisation (a), heating design, cooling design, and (b), simulation, computational fluid dynamics and daylighting using Radiance.

In addition to energy consumption calculations, the user can investigate the effects of natural ventilation and natural daylight, and carry out detailed HVAC design. The software is accredited for UK, France and Portugal Building Regulations and implements ASHRAE 90.1 energy models. It contains numerous component and template libraries, including information about building codes and weather data in numerous countries.

A very interesting tool in DesignBuilder is called 'People Payback Calculator'. This is particularly suitable for analysis of retrofit in non-domestic buildings, as it calculates return on sustainable investment resulting from retrofit interventions, taking into consideration sickness absence, productivity and staff retention.

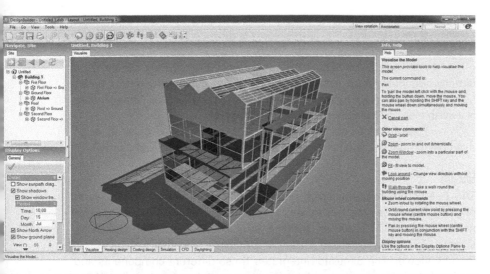

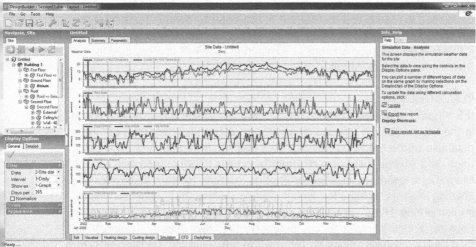

Figure 11.2 DesignBuilder. (a) Rendered geometry; (b) simulation results

Source: With permission from Design Builder Software Ltd

Energy Plus

Energy Plus (LBL-SRG, 2013) is a simulation engine with a long history, going back to the 1970s. It emerged as a consolidation of previous modelling software programs, including BLAST and DOE-2. It has the most comprehensive list of heat transfer and systems features. As it has been developed with US public funding, the user interface development has been left to the discretion of third parties.

Energy Plus is an open system under constant development. The simulation community is encouraged to make contributions in terms of new components of its open framework. This can be done without the knowledge of the overall system, thus enabling the contributors to focus on the functionality of new features. The core system is developed in FORTRAN, and it is supported by extensive documentation.

Although it is a simulation engine only, Energy Plus can be used directly, without a third-party user interface. Its native interface is basic; however, it provides a very powerful and useful simulation tool. The software is available on an Open Source License Agreement.

DesignBuilder software, described earlier in this chapter, is an example of how this simulation engine can be used by third-party software.

EDSL TAS

TAS is a commercial simulation product, which combines the simulation engine and the user interface in one package (EDSL, 2013). The core components include a 3D Modeller, Building Simulator, Construction Database, Internal Conditions Database, Calendar Database, Weather Database and Results Viewer. The Simulator integrates these different components and enables interactive access to them, using drag and drop functionality. TAS is developed in-house, and therefore it is not open to contributions from the simulation community. However, its interactive and intuitive interface makes it a very useful and comprehensive simulation tool. TAS is accredited for UK Building Regulations compliance.

IES Virtual Environment

IES Virtual Environment (IES, 2013) is commercial software that originated in the UK in 1990s. It is based on a number of modules that use a common database of the modelled building. This enables information to be entered once and shared between different modules. These modules include ModelIT for geometry creation or import; SunCast for solar geometry and shading analysis; Apache for dynamic thermal simulation; ApacheHVAC for implementing HVAC simulation; MacroFlo for bulk air flow movement between zones; MicroFlo for computational fluid dynamics simulation; Vista for results analysis; FlucsDL for daylighting analysis; RadianceIES for daylighting and electric lighting simulation; and a number of others that are beyond the scope of this text. IES is accredited for UK Building Regulations.

TRNSYS

The development of TRNSYS (TESS, 2013) has been continuous since the 1970s. Although it is based on FORTRAN, which is not an object-oriented language, the structure of TRNSYS is perhaps one of the first examples of a modular, object-oriented system. Initially it was developed at the University of Wisconsin to model solar energy systems, but its scope has been expanded considerably, to include modelling of buildings. The modular structure of TRNSYS lends itself to development contributions from the simulation community. The software documentation gives details of how individuals can develop new modules, thus making this an open and continuously expanding framework. TRNSYS is available on a commercial licence.

Steady-state calculation tools

Steady-state calculation tools employ low-granularity methods to estimate building energy performance. SBEM (BRE, 2013) and PHPP – Passive House Planning Package (Feist, 2012) are two examples of such tools. Calculations in these tools are carried out on the basis of monthly average values, and without using dynamic heat transfer principles. In comparison, dynamic tools, such as DesignBuilder, IES, TAS, and others, perform transient heat transfer calculations, at a minimum time step of one hour or shorter. We can therefore say that steady-state monthly average tools use twelve sets of numbers and dynamic tools use 8,760 sets of numbers, the latter corresponding to the number of hours in the year. If we put this as an analogy into the context of computer screen resolution, for instance, it is not hard to understand that a screen with 8,760 pixels will be far better than the screen with twelve pixels. Taking the analogy back into building simulation context, the twelve-'pixel' resolution tools have one more fundamental disadvantage, namely that steady-state heat transfer, which is the basis of these tools, does not occur in nature. The steady-state tools are therefore not suitable for design, and should not be used in the design decision-making.

In a recent comparative analysis of the performance of a building using PHPP and IES VE, PHPP did not predict any overheating in summer, whilst IES reported considerable overheating and consequent discomfort. This is just one example of why steady-state monthly average tools should not be used for design purposes.

Comfort modelling

Danish scientist Povl Ole Fanger (1970) found through experimental research that the perception of building occupants about their thermal environment depends on six variables: air temperature; mean radiant temperature; relative humidity; air velocity; resistance of clothing and personal metabolic rate. This can be expressed with the following equation:

$$PMV = f(T_a, T_{mr}, rh, v, M, R_c) \tag{3}$$

where

PMV	=	Predicted Mean Vote on a seven-point scale, where -3 means very cold, 0 means just right, and +3 means very hot.
f	=	A function of . . .
T_a	=	Air temperature
T_{mr}	=	Mean radiant temperature (a sum of products of temperatures of all surrounding surfaces and corresponding surface areas, divided by the total surface area of all surrounding surfaces)
rh	=	Air relative humidity
v	=	Air velocity

M = Metabolic rate, in Watts per surface area of the human body, expressed in units of 'met' (1 met = 58 W/m^2)

R_c = Resistance of clothing, expressed in units of 'clo' (1 clo = 0.155 m^{2}K/W)

Fanger established that a Predicted Percentage of Dissatisfied building occupants (PPD) is a function of the PMV, and that it can be expressed as a curve shown in Figure 11.3. As it can be seen from this figure, for the thermally neutral vote of PMV = 0, there are 5 per cent of dissatisfied people in the building. Therefore, even if we get our design absolutely right, there will be 5 per cent of dissatisfied building users. If the design results in a PMV that is slightly offset from the thermal neutrality, the number of dissatisfied building users will rise exponentially. Designers need to ensure that this number does not exceed 10 per cent PPD.

Most simulation software packages use the results of Fanger's research to establish thermal comfort parameters. For instance, the two personal comfort settings in IES VE (IES, 2013), namely the metabolic rate and the resistance of clothing, are adjustable only post-simulation, in the results review. Even here, these parameters can only be changed into one fixed value for the entire year, not taking into account hourly changes. This can lead to considerable over- and underestimates of thermal comfort in a building.

In a recent study on a closer analysis of adaptation of building users to climate change, it was demonstrated how adjusting these values post-

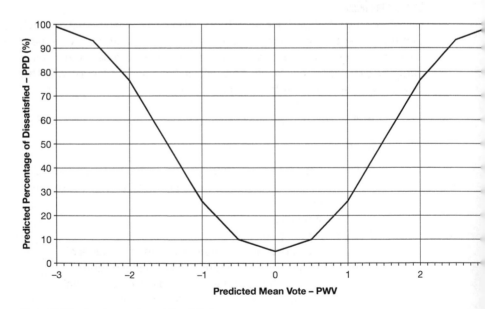

Figure 11.3 Predicted percentage of dissatisfied

Source: Ljubomir Jankovic

simulation in a standard spreadsheet programme on an hourly basis could lead to better designs that do not use mechanical cooling unnecessarily, as opposed to using the simulation results as they come out from the simulation model (Huws and Jankovic, 2013).

Computational fluid dynamics and comfort modelling

Results of dynamic simulations typically provide single values of parameters, such as air temperatures, per room and per time step, thus assuming that the distribution of that parameter is uniform throughout the volume.

However, distribution of air temperatures and other parameters that affect thermal comfort is rarely uniform in a three-dimensional space, and therefore a much more detailed analysis is required in order to investigate thermal comfort more accurately. This type of analysis is called Computational Fluid Dynamics, or CFD.

CFD is computationally very intensive. Whereas an annual dynamic simulation of a small commercial building, running in hourly time steps, could take a few minutes of computation, a CFD analysis for a single hour of the year could easily take more than an hour on a modern personal computer. Increasing the scale of the building will increase the computation time required to perform analysis.

Mainstream simulation software packages can perform simplified CFD analysis and produce outputs representing 3D distribution of temperature, velocity and pressure. In IES VE, the distribution of comfort indicators, such as the PMV and PPD, can be obtained by post-processing of the CFD results, after choosing activity and clothing levels and the methods for applying relative humidity and the mean radiant temperature in the comfort calculation. This effectively represents a decoupled thermal comfort model, which is produced merely as a post-calculation on the basis of CFD results. Examples of such CFD outputs for a small-scale commercial building are shown in Figure 11.4.

However, CFD and thermal comfort models influence each other, and hence the need for a coupled model for a more detailed analysis. Cook et al. (2011) used a breathing manikin with fifty-nine individual body segments based on IESD-Fiala model (IESD, 2012; Fiala et al., 1999) to investigate thermal comfort in a naturally ventilated classroom. The breathing function of the manikin was used for modelling CO_2 distribution and assessment of indoor air quality. One of the main findings was that this coupled model predicted lower PPD than the uncoupled model under increased internal temperatures. The main advantage of the coupled approach is in the prediction of comfort based on a detailed representation of the physiology and heat transfer on the surface of the human body, thus leading to more accurate results and more informed design decisions.

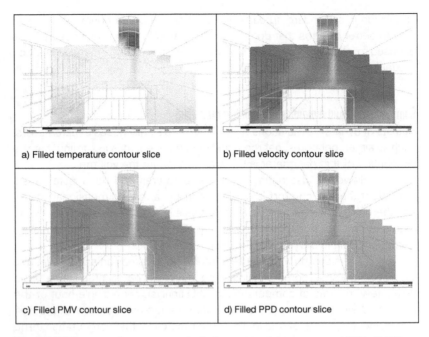

a) Filled temperature contour slice

b) Filled velocity contour slice

c) Filled PMV contour slice

d) Filled PPD contour slice

Figure 11.4 CFD-generated temperature and velocity contours and post-processed PMV and PPD
Source: (a) and (b) from Jankovic (2012a); (c) and (d) from Ljubomir Jankovic

USING ENERGY MODELS TO PREDICT LIKELY ENERGY USE IN PRACTICE

Absolute predictions by energy models can be inaccurate and can lead to a considerable discrepancy between the actual and modelled performance, called the performance gap. In the absence of measured performance that can be used for calibration of models and elimination of the performance gap, relative predictions that compare one variation of the model of the building with another can provide essential design information. We will now discuss the advantages and limitations of relative and absolute predictions and explain methods for elimination of the performance gap.

Modelling of alternative design solutions for the retrofit

Once we have the calibrated model, we can 'push it and poke it' and see what happens. This is not in any way a random process, but a simulation of a carefully selected number of design options that we would like to investigate by running the model.

Listing the design options will give us an idea of the size of the simulation job that we are embarking on, and a simulation plan. For instance, let us assume that we wish to investigate:

1   two different thermal insulation thicknesses;
2   two insulation placements (outside or inside);

3   two different window specifications (double- or triple-glazed);
4   two different heating systems (biomass or air-source heat pump);
5   three construction types (light, medium, high);
6   with or without free cooling.

As each of these options needs to be investigated with each other option, the size of the task is therefore 2 x 2 x 2 x 2 x 3 x 2 = 96 simulations.

The preparation of ninety-six versions of the same simulation model, running the model one at a time or as a batch process, and the interpretation of results could take quite some time.

With a large number of parameters, the results of correspondingly large number of simulations will be difficult for a designer to consolidate and find the best possible combination of design options.

However, recent developments of simulation tools include optimisation programs, which perform a number of parametric simulations, and optimise the candidate-solution space using evolutionary programming, such as genetic algorithms. These optimisation methods can save considerable time to the designer, by taking over the simulation process, searching the candidate-solution space, and finding the optimum set of design parameters. These new developments are discussed later in this chapter.

Performance gap and how to overcome it

The performance gap is well documented in the literature. In addition to the model-related inaccuracies caused by poor assumptions and the use of default parameter values in the absence of more realistic information from the building in use, as discussed by Menezes *et al.* (2012), there are also building-related causes of the performance gap. These are grouped into several categories of discrepancies, such as regulated energy (from fixed building services); unregulated energy (from lifts, security, plug loads, etc.); occupancy discrepancies between the model and the actual building; poor control, commissioning and maintenance, causing operational inefficiencies; and special functions, such as server rooms, trading floors and others (RIBA/CIBSE, 2010). The discrepancy between simulated and actual performance parameters can differ by as much as a factor of three, meaning that the simulation model can over- or underestimate by 300 per cent.

Even though the model of the building before retrofit may have been calibrated, the performance gap will still occur, albeit on a smaller scale, in relation to the design changes in the model. The elimination of the performance gap will therefore still be an issue and thus the simulation design analysis will need to be approached carefully.

One of the ways of overcoming the influence of the performance gap on the design decision-making is to use relative rather than absolute predictions, as explained in the next section. Another way would be to use advanced numerical methods that are still in the experimentation stage, and these can lead to advances in modelling and simulation. One of these methods is explained in the section after next.

Relative versus absolute predictions

An absolute prediction from a simulation model is a simulation result that is taken in absolute terms. In other words, if the result of simulation tells us that the total annual energy consumption is X MWh per annum, then we use that result in our design decision-making.

From experience of numerous users of building simulation software, and as discussed in the previous section, this approach can result in incorrect design decisions, and will certainly cause the performance gap, with considerable consequences on the actual energy consumption and carbon emissions in a retrofitted building.

A way to reduce the effect of the performance gap is to take relative comparisons between simulations. For instance, if we take the base line model of the building before the retrofit, change one design parameter, and compare the results of simulations before and after changing that parameter, the comparison will give us a relative prediction of the performance of the corresponding design decision. Instead of describing the building performance after the retrofit in absolute terms, such as X MWh of energy consumption per annum, we can now describe it as Y per cent better (or worse) than the base case. This will enable the designer to make informed decisions about choices of design parameters.

When making relative predictions of building performance through simulation, it is critical to change only one design parameter at a time. This ensures that the effect of that parameter alone is investigated, and its design merits are properly assessed, without influences from other parameters.

Reducing the performance gap using Fourier filters

A method for reducing the performance gap has been developed using principles of digital signal processing based on Fourier series, as published at the IBIPSA 2013 Conference (Jankovic, 2013). Fourier transforms of results of a calibrated and uncalibrated simulation model are first created using internal air temperatures in a free-floating mode, and the ratio between the two transforms, called Fourier filter, is created (Figure 11.5). When this filter is applied to simulation results of a sufficiently similar non-existing building, it will morph these results into values that are equivalent to the results of a calibrated simulation model of the non-existing building.

In this way, the Fourier filter captures the performance gap in the existing building, and eliminates it in the building on a 'drawing board'. This method is yet to be verified in practice through monitoring, simulation and numerical analysis; however, the initial results are encouraging, and they can lead to the development of dynamic benchmarks of building performance.

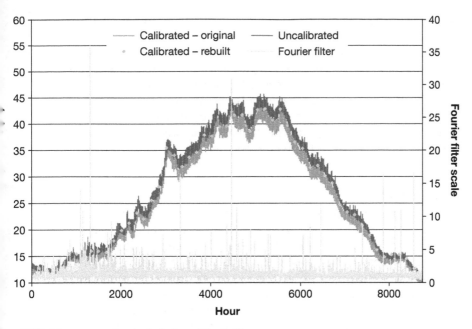

ure 11.5 Building response function in the form of Fourier filter

urce: Adapted from Jankovic (2013)

Unlike the improved conventional benchmarks of energy perform-ance, such as those proposed by Menezes *et al.* (2012), which comprise of a single number in terms of kWh/m^2/year, the Fourier filter-based bench-marks would capture the dynamics of building performance on an hourly basis over the entire year and could be applied to the simulation results that include the performance gap, and morph them into the results in which the performance gap is eliminated.

The question then arises as to how these simulation results, which have been morphed through post-processing outside the simulation model, could help the designer to gain the insight into which parameters in the simulation model caused the performance gap in the first place?

The answer is simple: the morphed simulation results can be used as a target for the calibration of the simulation model of a building on the drawing board. After the calibration, the designer can experiment with a higher accuracy simulation model, and use it in the design decision-making.

FUTURE DEVELOPMENTS IN REALISTIC MODELLING

The state of the art of today's modelling is heavily influenced by the development of science and mathematics over the past 300 years, and has resulted in limitations of the current models and in the accuracy of their predictions. In this section we will discuss new developments in modelling, which could lead to more accurate representations of buildings and reduced model development and simulation time. This will include: optimisation models that use evolutionary computing to search candidate-solution space; Fourier filters that can lead to simplified but accurate models of buildings; emergence-based approaches that give rise to self-organising models; and Building Information Models that integrate architectural, structural and systems aspects of buildings, but need to undergo further development in order to integrate energy and comfort modelling.

Optimisation models

Design based on modelling and simulation often involves investigation of the effect of different design options. As explained earlier in this chapter, a relatively small number of design parameters can increase the number of simulations exponentially. The designer then needs to search through the numerous simulation results files and select the optimum set of design parameters on the basis of interpretation of these results.

Optimisation methods can help considerably with automating this process, using evolutionary computing algorithms. These methods can take into account both technical aspects in terms of CO_2 emissions and thermal comfort, as well as capital and running costs, and optimise the design solution.

There were three projects that were developing optimisation methods at the time of writing of this chapter. These were Optimise Project (OPTIMISE, 2013); jEPlus (Zhang, 2012); and Advanced Design + OPTimisation (ADOPT, 2013). There have been several other developments in this area over the past decade, but for various reasons they have not become available for public use.

Optimise Project uses IES Virtual Environment software (IES, 2013), and applies the NSGA-II, a multi-objective genetic algorithm for searching the space of candidate solutions (Deb *et al.*, 2002). Descriptions, available from on- and offline sources, explain that this software will enable the user to make trade-offs between conflicting constraints in the design of a building, whilst using predetermined design parameters as variables.

jEPlus (Zhang, 2009) started as an EnergyPlus simulation manager for parametric runs. A set of user selected parameters are varied within a predetermined range, and each set is inserted into EnergyPlus to conduct a corresponding parametric simulation. Results from all simulations can be reviewed using external graphics packages. Although jEPlus itself does not perform optimisation, Zhang (2012) used it successfully with external evolutionary algorithm packages to conduct single- and multi-objective optimisation of heating and cooling energy consumption in an office building. In a subsequent development, jEPlus is used in a project named ADOPT – Advanced Design + OPTimisation (ADOPT, 2013), which uses DesignBuilder software (DBS, 2013) as the host application, and which will be capable of performing optimisation, whilst supporting multiple user interfaces and providing an online service.

Genetic algorithms, used in the above developments, are nature inspired. Each parametric case in the building simulation is represented with a long character string, equivalent to a biological gene, and has an assigned fitness function. The collection of character strings representing different simulation cases is then subjected to genetic operators: mutation, crossover and reproduction. The latter operator carries out the selection of the fittest genes (simulation cases) using the fitness function. In this way the optimisation occurs on the basis of the parameter sets within a population of strings that represent the space of candidate solutions. This enables the genetic algorithm to search the candidate solution space relatively quickly, taking into account multiple objectives, and presenting the designer with considerably narrowed down design solutions and trade-offs between these solutions. An excellent introduction into genetic algorithms was published by Goldberg (1989), and to this day it remains one of the best and most general texts on the subject.

Although the building simulation optimisation software packages discussed here were not available for public use at the time of writing of this chapter, early indications are that they could considerably change the culture of building simulation as we know it today, by searching large-candidate solution spaces and finding solutions that designers could not search very easily using conventional methods.

Models based on Fourier filters

Earlier in this chapter we explained that a Fourier filter was obtained as a ratio between Fourier transforms of results of a calibrated and uncalibrated simulation model, which captures the dynamic relationship between the two.

The same method can be used for the creation of a simplified simulation model of a building. To start, we need to have results from monitoring of that building, including external driving functions, such as air temperature and solar radiation, as well as internal air temperatures. In the first step, a combined influence of external temperature and solar radiation is created, as shown in Equation 4:

$$t_{s-a} = t_o + \frac{\alpha(I_{dir} - I_{dif})}{h_s} \tag{4}$$

where

t_{s-a} = sol-air temperature (°C)
t_o = outside air temperature (°C)
α = external wall absorption coefficient (dimensionless)
I_{dir} = direct solar radiation perpendicular on the wall surface (W/m^2)
I_{dif} = diffuse solar radiation (W/m^2)
h_s = outside surface conductance (W/(m^2.°C)).

In the next step, Fourier transforms of a representative internal air temperature and the sol-air temperature are created. A ratio between the two Fourier transforms comprises a Fourier filter, which in fact represents a compact model of the building in the form of a response function. If we apply the filter on the Fourier transform of the same sol-air temperature, we will get the internal air temperature. Conversely, if we apply the Fourier filter on the Fourier transform of another sol-air temperature, such as from a future climate year, we will get the internal air temperature of the building influenced by the future climate. Experimentation with this approach to simplified modelling has led to promising results.

This is just one example of how Fourier transform and Fourier filters can be used for simplified but accurate building simulation models. The method requires more work in order to achieve creation of accurate models from short-term, rather than from annual, monitoring but nevertheless this gives an insight into the future possibilities for obtaining simplified but accurate simulation models.

Emergent self-organising models

Buildings do not know how to carry out numerous calculations in order to transmit heat; just as bridges do not know how to perform calculations of structural dynamics in order to decide whether to stand or to collapse. These processes in man-made structures occur seamlessly, just as they occur in natural systems. For instance, a DNA molecule does not perform billions of computations in order to form itself, but it simply self-organises into a particular double-helix structure that is the essence of life on the planet.

This approach to modelling was brought to a larger audience by the work of Craig Reynolds on self-organising models of animal group formations (Reynolds, 1987). He showed that using three simple rules, a realistic model of a flock of artificial birds could be created in the computer model. As it was roaming around the artificial terrain, the flock split around the obstacles and re-joined spontaneously, without explicit instructions to do so.

Similar principles of emergence and self-organisation were applied to modelling of engineering structures, such as bridges and domes, as well as to modelling of self-organised building forms (Jankovic, 2012b).

Following the same logic, it will not be impossible to develop emergent and interactive models of CFD, which would operate on a timescale of seconds, rather than hours, in order to calculate air temperatures and velocities in the entire volume of a room. Instead of performing an annual simulation of a building based on hourly time step calculations of numerous equations as conducted by today's simulation models, an hourly time step emergent CFD model that takes a few minutes to complete an annual simulation would represent a true culture change in this field.

Building Information Modelling

Building Information Modelling (BIM) refers to an integrated model, in which components, such as an architectural model, a structural model and an M&E model, all use the same database, so that the design generated by each component works interactively with designs from each other component (ARUP, 2013). This, for instance, enables early collision detection between mechanical and electrical services on the one hand and the architectural and structural design on the other. It also enables the sizing of ducts and pipes in the context of architectural or structural aspects of the building design. Through special add-on hardware and software, a BIM design can be 'played back' at the building site, determining the construction sequence down to fine detail, including, for instance, the location and specification of individual holes that need to be drilled, pointed to by a laser playback device.

At the time of writing, energy and comfort modelling was not fully integrated into BIM systems. Mitchell (2011) explained the challenges and opportunities of such integration. Considering the current momentum in the research and development in that area, it is expected that BIM could become a predominant method for design, construction, management and post-occupancy evaluation of buildings, to include fully integrated energy and comfort modelling.

CONCLUSIONS

The simulation model is a comprehensive response function that takes external input, applies internal transformation processes and creates output. As simulation models can exhibit a considerable performance gap if they are not calibrated, it is especially important that all relevant information about the building is as accurate as possible, and that calibration of the model is carried out using measured performance data from the actual building. This sensitivity to the quality of inputs is often described as GIGO – Garbage In, Garbage Out, and it is a useful reality check for everyone involved in building simulation.

It is important to differentiate between the dynamic simulation models on the one hand and the steady-state calculation tools on the other. The former are high-granularity tools that implement dynamic heat transfer principles and perform calculations on an hourly basis, and the latter are low-granularity tools that perform simplified calculations on a monthly average basis. Considering that even the high-granularity dynamic tools can lead to the performance gap and design errors, the simplified tools have no place in the design decision-making, but can be and are used for compliance checking.

Once we have a calibrated simulation model, we can 'push it and poke it' with different design and input parameters. This will enable us to investigate various aspects of the design, and to make informed decisions based on the model performance.

The essential strategy for robust design decision-making is to use relative comparisons between simulations of different designs, and not to rely on absolute results of individual simulations, thus reducing the effect of the performance gap.

Building simulation can be considered to be an art, analogous to playing the violin: the more one practices, the better it goes, especially if one has a bit of talent. However, future developments of simulation models are going to make this easier for the user. Developments of optimisation methods will help the user search large candidate-solution spaces automatically, using genetic algorithms running in the background. Nature-inspired emergent modelling as well as modelling with response functions in the form of Fourier filters will produce accurate answers with a considerably lower user effort. BIM will develop beyond its current state of the art and it will fully integrate the advances in energy and comfort modelling, so as to create the building in parallel reality, and one-to-one correspondence between the actual building and its model, where changes in one are automatically reflected in the other and vice versa. The future developments in realistic modelling are beginning to look so exciting that it will be hard to wait.

REFERENCES

ADOPT (2013) *Advanced Design + OPTimisation*. Available from: www.iesd.dmu.ac. uk/~adopt/wiki/doku.php (accessed 1 May 2013).
ARUP (2013) *Building Information Modelling (BIM)*. Available from www.arup.com/ Services/Building_Modelling.aspx (accessed 1 June 2013).
BRE (2013) *SBEM: Simplified Building Energy Model*, Building Research Establishment, Garston. Available from: www.bre.co.uk/page.jsp?id=706 (accessed 1 May 2013).
CIBSE (2006) *TM22: Energy Assessment and Reporting Method*, London: Chartered Institution of Building Services Engineers.
Crawley, D. B. (1998) 'Which weather data should you use for energy simulations of commercial buildings?', *ASHRAE 1998 Transactions*, 104(2): 498-515.
Cook, M., Yang, T. and Cropper, P. (2011) 'Thermal comfort in naturally ventilated classrooms: application of coupled simulation models,' *Proceedings of Building Simulation 2011: 12th Conference of International Building Performance Simulation Association*, Sydney, 14–16 November. IBPSA. Available from: www.ibpsa.org/ proceedings/BS2011/P_1714.pdf (accessed 1 May 2014).
DBS (2013) *DesignBuilder*, Design Builder Software Ltd. Available from: www.design builder.co.uk/ (accessed 1 May 2013).
Deb, K., Pratap, A., Agarwal, S. and Meyarivan, T. (2002) 'A fast and elitist multiobjective genetic algorithm: NSGA-II', *IEEE Transactions on Evolutionary Computation*, 6(2): 182–97.
EDSL (2013) *TAS Building Designer*, Environmental Design Solutions Ltd. Available from: www.edsl.net (accessed 1 May 2013).
Goldberg, D. E. (1989) *Genetic Algorithms in Search, Optimization & Machine Learning*, Reading, MA: Addison-Wesley Publishing Co.
Fanger, P. O. (1970) *Thermal Comfort*, Copenhagen: Danish Technical Press.
Feist, W. (2012) *The Passive House Planning Package (PHPP)*, Passivhaus Institut. Available from: www.brebookshop.com/details.jsp?id=326953 (accessed 1 May 2013).
Fiala, D., Lomas, K. J. and Stohrer, M. (1999) 'A computer model of human thermoregulation for a wide range of environmental conditions: The passive system', *Journal of Applied Physiology*, 87(5): 1957–72.

Huws, H. and Jankovic, L. (2013) 'Implications of climate change and occupant behaviour on future energy demand in a Zero Carbon House', *Proceedings of the 13th International Conference of the International Building Performance Simulation Association*, 25–28 August 2013, Chambery, France. IBPSA. Available from: www.ibpsa.org/proceedings/BS2011/P_1714.pdf (accessed 1 May 2014).

IES (2013) *Virtual Environment 2012*, Integrated Environmental Solutions Ltd. Available from: www.iesve.com (accessed 1 May 2013).

IESD (2012) *IESD-Fiala Model*. Available from: www.iesd.dmu.ac.uk/~yzhang/wiki/doku.php?id=software:model:iesd_fiala:introduction (accessed 1 June 2013).

Jankovic, L. (2012a) *Designing Zero Carbon Buildings Using Dynamic Simulation Methods*, London and New York: Routledge.

Jankovic, L. (2012b), 'An Emergence-based Approach to Designing', The Design Journal, Volume 15 Issue 3, pp. 325-346, September 2012, London: Berg.

Jankovic, L. (2013) 'A method for reducing simulation performance gap using Fourier filtering', *Proceedings of the 13th International Conference of the International Building Performance Simulation Association*, 25–28 August 2013, Chambery, France. IBPSA. Available from: www.ibpsa.org/proceedings/BS2011/P_1714.pdf (accessed 1 May 2014).

Jankovic, L. and Huws, H. (2012) *Simulation Experiments with Birmingham Zero Carbon House and Optimisation in the Context of Climate Change in England*, Building Simulation and Optimisation 2012, Loughborough: IBPSA.

LBL-SRG (2013) *EnergyPlus Energy Simulation Software*, Lawrence Berkeley Lab - Simulation Research Group. Available from: http://apps1.eere.energy.gov/buildings/energyplus/ (accessed 1 May 2013).

Menezes, A. C., Cripps, A., Bouchlaghem, D. and Buswell, R. (2012) 'Predicted vs. actual energy performance of non-domestic buildings: Using post occupancy evaluation data to reduce the performance gap', *Applied Energy*, 97(September): 355–64.

Mitchell, J. (2011) 'BIM & building simulation', *Proceedings of Building Simulation 2011: 12th Conference of International Building Performance Simulation Association*, Sydney, 14–16 November. IBPSA. Available from: www.ibpsa.org/proceedings/BS2011/P_1714.pdf (accessed 1 May 2014).

OPTIMISE (2013) *Optimise Project*. Available from: www.optimise-project.org (accessed 1 May 2013).

Reynolds, C. W. (1987) 'Flocks, herds, and schools: A distributed behavioral model', *Computer Graphics*, 21(4): 25–34.

RIBA/CIBSE (2010) *Carbon Buzz*. Available from: www.carbonbuzz.org/docs/Carbon Buzz_Handbook.pdf (accessed 1 April 2013).

TESS (2013) *TRNSYS – Transient System Simulation Tool*, Thermal Energy System Specialists, LLC. Available from: www.trnsys.com (accessed 1 May 2013).

US DOE (2013a) *Building Energy Software Tools Directory*, US Department of Energy. Available from: http://apps1.eere.energy.gov/buildings/tools_directory/ (accessed 1 May 2013).

US DOE (2013b) *Weather Data Information*. Available from: http://apps1.eere.energy.gov/buildings/energyplus/pdfs/weatherdatainformation.pdf (accessed 1 April 2013).

Zhang, Y. (2009) '"Parallel" EnergyPlus and the development of a parametric analysis tool', *IBPSA BS2009*, Glasgow, 27–30 July. IBPSA. Available from: www.ibpsa.org/proceedings/BS2011/P_1714.pdf (accessed 1 May 2014).

Zhang, Y. (2012) 'Use jEPlus as an efficient building design optimisation tool', *CIBSE ASHRAE Technical Symposium*, Imperial College, London, 18–19 April. IBPSA. Available from: www.ibpsa.org/proceedings/BS2011/P_1714.pdf (accessed 1 May 2014).

12
WATER, WASTE, MATERIALS AND LANDSCAPE
Paul Appleby

INTRODUCTION

Sustainable refurbishment is not only about reducing operational energy and associated CO_2 emissions and improving comfort and indoor environmental quality. Opportunities should also be taken to incorporate other sustainability measures such as reducing water consumption and improving management of occupant generated waste. Refurbishment is inherently more efficient in its use of materials than providing the same accommodation by new build; whilst any new materials used can be specified with low environmental impacts and embodied energy. Refurbishment may also enable incorporation of enhanced ecological features and sustainable surface water drainage.

All of the above are associated with environmental impacts, including the whole-life greenhouse gas emissions associated with a refurbished building or development. These include, for example, the treatment and distribution of water and sewage, the collection and processing of waste, the methane emissions from decomposing wastes and the energy associated with manufacture, delivery and eventual demolition and disposal of building materials. As well as their impact on local biodiversity, decisions on landscape refurbishment may also impact carbon emissions associated with adsorption and emission of CO_2 by trees and other vegetation.

This chapter sets out a distillation of the key design requirements when incorporating the above into the refurbishment of a commercial building located in a temperate climate zone. This will include:

- water saving measures such as retrofitting water-efficient appliances, rainwater harvesting and grey water recycling, and on-site water treatment;
- retrofitting integrated waste management: facilities for recycling;

- landscaping, sustainable drainage and biodiversity: invasive species, enhancing ecological value, green roofs and walls, swales, permeable paving;
- materials and embodied energy/carbon: reducing materials impacts and waste.

For a more detailed account of these measures and associated legislative and planning context refer to *Sustainable Retrofit and Facilities Management* (Appleby, 2013).

WATER SAVING

Background

Water is a valuable resource and although there are some regions that have more than they need, long periods of drought impact a large proportion of the world population. The developed world in particular is profligate with the resources they have, with nations such as the USA and Australia having daily per capita abstraction rates of 4,740 and 3,562 litres per person respectively, compared with 630 litres for the UK.[1] Estimates for developing countries vary from 50 to100 litres/person/day.

In the UK most authorities agree that office water consumption averages at between 16 and 50 litres/worker/day, depending on whether there is a canteen, showers, humidifiers, evaporative cooling towers and soft landscaping.[2] Cooling towers and humidifiers might represent some 25–35 per cent of water use whilst, where there are water-hungry plants, irrigation can represent up to 38 per cent of demand. Where there is no demand from cooling tower make-up or landscaping then toilet flushing tends to dominate, followed by urinal flushing, hand washing, canteen use and office cleaning. Leakage can represent a significant draw on the water supply, particularly for more complex and older sites.

Hospitals, sports centres, hotels, garden centres, residential care homes, car wash facilities and certain industrial processes tend to have the greatest intensity of water consumption, particularly where there are swimming pools, spa pools and water is used for washing, air cleaning, manufacture, therapy or treatment.

The process of abstracting and storing water, adding chemicals, distributing it to consumers then handling, processing and discharging or recycling the sewage uses primary energy, resulting in associated carbon emissions, whilst also having environmental impacts associated with sewage management and chemical water treatment. Although chemicals that have significant impact in high enough concentrations, such as chlorine, are used in water treatment, the most significant impacts come from sewage settlement and treatment, with associated handling of liquid, slurry, sludge and solid residuals. Odour comes from the occasional overflow of untreated sewage as well as the formation of acetic acid and methyl acetate during settlement.

Water saving measures

The extent to which water consumption can be reduced through refurbishment will depend on the efficiency of existing water consuming appliances and systems and the extent of refurbishment planned. Where, for example, it is proposed to upgrade existing sanitary facilities it may be feasible to use the same approach as for new build. An understanding of existing water consumption and inefficiencies can be used to inform decisions on the extent of this upgrade. For example, if there is a history of leaking joints and washers there is a strong argument for complete replacement of pipework and appliances.

Sanitary appliances

Although dual-flush WCs have become the norm, their performance can be disappointing. In theory they should be used on low flush three times for every single full-flush operation, resulting in an average water use of 4 litres per flush for a cistern having a flush ratio of 6:3 from full to reduced (i.e. a 33 per cent reduction). However studies in a residential context have found that water consumption only dropped by 18 per cent of full-flush volume compared with a 6-litre cistern.

For non-residential buildings where urinals are provided there is little point in installing dual-flush cisterns because the WCs will tend only to be used for solids.

Bearing in mind the risk of leakage through standard valve cisterns the designer should consider retrofitting dual-flush or controllable siphon devices along with delayed-action inlet valves.

Older urinal installations might typically have a continuous flow of water, consuming perhaps 500 m^3 of water a year to flush four or five stalls, regardless of use. In fact the UK's Water Supply Regulations stipulate that each stall should be flushed with no more than 7.5 litres per hour per stall. However, rather than using a manually adjusted valve to allow a continual flow at this rate, a variety of low-cost automatic systems are available, including solenoid valves operated from timers, door switches, infrared (IR) presence detector or water pressure fluctuation when a hot tap is operated. For maximum economy, flushing for individual stalls can be provided using IR devices or a spring-loaded push button or lever-operated valve.

Alternatively, where a major refurbishment of sanitary accommodation is planned, existing stalls can be replaced with waterless urinals. These incorporate a deodorising chemical seal that is lighter than urine and caps the outlet thus masking the odour from the drain (see Figure 12.1).

Cross-Section of the Patented Vertical EcoTrap®

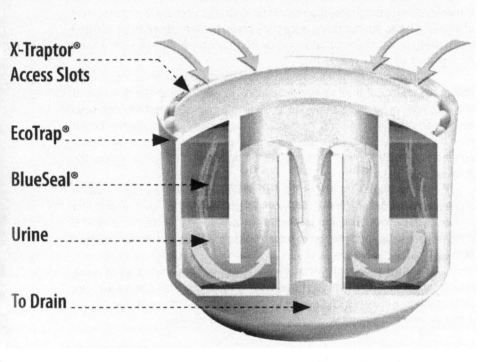

X-Traptor®
Access Slots

EcoTrap®

BlueSeal®

Urine

To Drain

Figure 12.1 Waterless urinal: typical outlet showing operating principle
Source: Waterless Co

Taps

The type of tap that offers the greatest water saving will depend on how it is to be used. A tap over a kitchen sink or bath will usually be used for filling things and hence the flow rate is probably not very important. Taps used primarily for washing hands, however, can perform well with low flow rates provided their wetting capability is maximised. Hence aerated or spray taps are ideally suited to hand washing and can save up to 80 per cent of water consumption compared with a conventional bib tap.

Products are available that operate as a spray tap when the handle is partially turned then open to full bore for basin filling when operation is continued. Where there is a risk of taps being left running, which applies to most taps used for hand washing, then a degree of automation is worth considering. This can be achieved with IR operation or self-closing push-button taps, although the reliability of the latter is varied, with large variations in delay before closing possible. IR has the advantage of hygienic operation, although batteries or electricity supply will be required.

Showers

Showers installed in non-residential buildings tend to be used by those who have undertaken athletic activities such as cycling, swimming or running. Generally time spent in the shower will be shorter, but there is a greater risk of showers being left operating once the person has left. This can be avoided by installing timed push-button valves or IR presence detection, with a fixed temperature setting.

Although electric showers are the most popular choice for retro-fitting, the quality of the shower experience is variable and the duration of showers may need to be longer because of the time required to rinse soap away, particularly from long hair. Aerated shower heads achieve better wetting at similar flow rates because the spray is mixed with air. Some aerated shower heads are advertised as being suitable for use with electric showers.

According to the UK Water Technology List: 'An efficient aerated showerhead is defined as a showerhead that mixes air and water and delivers a fully formed spray pattern, with a flow rate of no more than 9 litres/minute when operated at dynamic pressures up to 5 bar (for all spray settings).'[3]

Another source of wastage associated with showering is the water run off whilst adjusting temperatures with a mixer tap. This can be avoided by retrofitting a mixer having separate thermostatic adjustment that can be left at the desired setting whilst the water valve is opened.

Washing machines and dishwashing

To be listed in the UK's Water Technology List and qualify for enhanced capital allowances (ECA) commercial washing machines must be of the horizontal axis type 'with an energy and wash performance equivalent to the European Energy Label rating of AA for energy consumption and wash performance respectively. The machine must also not exceed a maximum water consumption of 12 l/kg wash load'.[4] The list also covers industrial washing machines such as batch washers, washer tunnels and washer extractors.

> Typically, large restaurants and food service operations utilize commercial dishwashers. Prior to loading the dishwasher, plates and dishes are manually sprayed (pre-rinsed) to remove loose or 'sticky' food. The washing of dishes typically consumes two-thirds of all water use from the restaurant. The water used in this pre-rinsing operation is often twice the volume of water used by the dishwashing equipment. The most cost-effective water conservation measure in a commercial food service operation is improving the efficiency of the pre-rinse spray valve.[5]

The Water Technology List does not cover dishwashers but the UK 'Government Buying Standards' site refers to the EU Energy Label scheme[6] for domestic scale appliances and the US Energy Star specification[7] for larger commercial dishwashers. For the former, the mandatory standard is for a minimum A rating covering energy, cleaning and drying with water consumption of 12 litres for fourteen place settings, dropping to 10 litres for a best practice standard.

Evaporative cooling towers and condensers

The amount of water used in an open cooling tower depends on how much is lost through drift, evaporation, leakage and windage/splashing from the water that cascades through the pack or heat exchanger into the open pond at its base, along with the amount of 'blowdown' or bleed required to control the concentration of salts, such as chlorides, and other dissolved impurities (total dissolved solids – TDS), in the circulating water. Evaporative loss depends on the amount of heat rejected, whilst other losses depend on the spray generated and the amount captured in the drift eliminators located normally at the tower discharge and the splash guards located at the air inlets. Modern towers are designed with very effective drift eliminators to reduce the risk of *Legionella* bacteria being carried out of the tower in very fine water droplets. Typically between 2.8 and 3 litres/h of make-up water will be required per kW of heat rejection capacity. Evaporation losses depend on the water circulation rate and the difference between the water temperature and the dew point temperature of the air. Hence as the amount of heat rejected reduces the evaporation rate should fall, assuming water is diverted from circulating through the tower as demand reduces. A cooling tower serving an office building having a significant cooling load might use around 800 litres/m²/annum. For a 10,000 m² office block this equates to 8,000 m³ of water per annum.

Although refurbishing or replacing an existing cooling tower will reduce water consumption the ideal solution would be to remove it and modify the system to operate with a dry tower or condenser. In some cases it may be possible to dispense with air conditioning altogether.

Humidifiers

The amount of water required for humidification depends on the amount of air to be humidified, the percentage saturation to which the building is controlled, the amount of fortuitous moisture gain to the treated space and the moisture content of the outside air. The latter value varies with time, whilst moisture emissions will vary with occupancy and other sources, such as cooking and kettle boiling.

Steam humidification requires a source of steam, either from a central boiler or from an electric or gas-fired steam generator, piped into a supply air duct via a header containing an array of nozzles. Not all of the water supplied ends up increasing room humidity: some is 'blown down' for the same reasons described above for evaporative cooling towers; whilst some is lost through condensation on surfaces within the ductwork. These factors can result in 50–100 per cent more water being drawn from the mains than is used for humidification.

When retrofitting humidification equipment it is worth considering devices that incorporate condensate recovery. As with cooling towers the ideal solution is to eliminate humidifiers altogether, based on assessment of need.

Retrofitting rainwater harvesting and grey water recycling

Rainwater harvesting systems for medium- to large-scale buildings typically comprise catch pits, filters, underground storage, water treatment and pump. These might use roof and surface water for toilet flushing, swimming pool or cooling tower feed, as well as irrigation.

The storage tank is normally sized from the amount of rainwater likely to be available, which depends on the surface area of the roof, its slope and type of roof surface. For example, typically only 50 per cent of rain that falls on a flat roof is available for harvesting (i.e. a drainage factor of 0.5). This proportion falls to even lower levels for green or landscaped roofs, which do not make an ideal catchment area for rainwater harvesting. Smooth sloping roofs can have a drainage factor as high as 0.9. The amount of water stored is usually calculated from 5 per cent of the total annual rainfall adjusted for the drainage factor of the roof and the filter efficiency of the installation (available from the manufacturer, but typically 0.9). Coarse filters are required to remove leaves and other large solids. Beyond this, the level of filtration and water treatment provided depends on the appliances being supplied. It may be considered that coarse filtration is sufficient for irrigation and even supplying WC cisterns, although there have been some problems with staining of WC basins from residues in the harvested water. Some installations have incorporated additional filtration and ultraviolet biocidal control, although the cost, energy consumption and additional maintenance from these must be factored into the feasibility study for the installation.

Water that has been used for washing purposes ('grey' water) can also be collected, cleaned to some extent and reused for flushing toilets and even irrigation and to supply washing machines, if the quality is good enough. Generally speaking water is collected from showers, baths and wash hand basins, and possibly from washing machines and dishwashers, although the cleaning system will have to be sufficiently robust to handle the additional detergent and grease loads from these appliances. Stored water is held for a time to allow sludge to collect and surface scum to be removed automatically, and both flushed away. If there is no demand for a predetermined period the whole system is drained and flushed through.

For larger installations a water treatment plant can be installed on similar lines to a municipal plant, but the additional carbon emissions associated with water treatment will have to be justified by the water savings achieved.

Both of these types of system require complete re-plumbing and space for storage, hence they would only be suitable for major refurbishments.

For detailed design guidance refer to the *Reclaimed Water* publication from CIBSE (2005) and advice and guidance from Anglian Water and downloadable from the UK rainwater harvesting association website.[8] Rainwater harvesting equipment appears in the Water Technology List[9] and hence qualifies for tax relief in the UK through enhanced capital allowances (ECA).

INTEGRATED WASTE MANAGEMENT

According to the UK Government's Department for the Environment, Food and Rural Affairs (DEFRA) 'the UK consumes natural resources at an unsustainable rate and contributes unnecessarily to climate change. Each year we generate approximately 290 million tonnes of waste, which causes environmental damage and costs businesses and consumers money.'[10]

Processing and landfill of waste have considerable environmental impacts, including greenhouse gas emissions associated with methane production at landfills and wastewater treatment plants. The US Environmental Protection Agency (EPA) reports that 161 million tonnes of CO_{2eq} were emitted from these sources in 2006 (EPA, 2008). This can be compared with 5,600 million tonnes associated with energy consumption (CO_2 only) (2008 figures from IEA, 2010).

When planning for refurbishment a review of the existing waste strategy should reveal whether there is sufficient provision for recycling of waste generated from the activities carried out within the building. Most councils and local authorities will have a collection strategy for recyclable waste, although these are reviewed from time to time when there are changes in legislation and budgetary constraints. Space needs to be allocated for suitably sized bins close to locations where waste is generated and where it is to be collected by the refuse collection vehicles.

The types of waste to be separated and the volumes generated will depend on the activities for which the building is intended and the policy of the local authority on grouping of recyclable wastes, which in turn depends on the design of the facilities to which the waste is sent.

LANDSCAPING, SUSTAINABLE DRAINAGE AND BIODIVERSITY

Whilst the construction of a new building frequently involves destroying habitats or disturbing ecosystems, refurbishment of an existing building may be possible without having a negative impact on the ecological value of soft-landscaped areas surrounding the building. This will depend on the amount of work required externally: building an extension, for example, may be as challenging as new build, whilst applying external treatments or erecting scaffolding may disturb or destroy vegetation close to the walls.

On the other hand refurbishment may provide an opportunity to enhance the ecological value of a previously barren landscape by judicious planting. Similarly there may be opportunities to reduce flood risk by increasing the permeability of the site and introducing sustainable drainage systems. Also it might be worth exploring the potential to create green roofs, brown roofs or green walls on the building envelope.

A major concern that must be addressed prior to restoring a degraded landscape is the identification and eradication of invasive species. Invasive alien species, such as the notoriously destructive Japanese knotweed and the phototoxic giant hogweed (see Figures 12.2 and 12.3), have become a major focus of international conservation concern and the subject of cooperative international efforts, such as the Global Invasive Species Programme (GISP). In the introduction to its landmark strategy document GISP states that 'Invasive alien species are now recognized as one of the greatest biological threats to our planet's environmental and economic well being.'

Figure 12.2 Japanese knotweed

Source: Gav

Figure 12.3 Giant hogweed
Source: Appaloosa

If invasive species are identified, an eradication method should be developed and implemented, using a specialist contractor. A number of eradication techniques are used with varying degrees of success. Critical to the complete eradication of Japanese knotweed, for example, is killing the rhizomes and the complete removal and disposal of all plant material, either by deep burial on-site or removal and disposal in a licensed site. In the UK, Japanese knotweed is defined as a controlled waste. The method employed will depend primarily on the length of time available for treatment. The Environment Agency has published *Managing Japanese Knotweed on Development Sites* which sets out the treatment options and the techniques available for management and eradication of the weed (EA, 2006).

Landscaping

While renovating a building it is important to preserve the existing trees and other landscaping and avoid unnecessary damage from the works. A survey of local trees and flora will show what can best be planted to enhance the local environment in terms of shading for pedestrians and ecological improvement. Landscaping in urban areas has a significant effect on issues such as the urban heat island effect and atmospheric pollution levels, and increasing the number of trees, shrubs and green areas around buildings will enhance the environment both for the occupants and generally in the surrounding areas.

Sustainable drainage

The efficient and rapid removal of storm water without overflow or back-up is important in mitigating the risk of flood, land pollution and soil erosion, with the related risks from undermining of foundations, damage to property and destruction of flora. The extent of impermeable surfaces in landscaped areas around a building directly impacts on the volume of storm water that has to be handled by the drains and hence flood risk. Where combined sewers are prevalent, such as across most of London, this can lead to discharge of untreated sewage into watercourses, or even backing up of sewage. Reducing the volume of water that is discharged into a combined sewer can also result in significant savings in sewer charges.

The above points are reinforced by the UK Environment Agency in their 2007 report on SUDS (Sustainable Urban Drainage Systems) retrofit in urban areas in which they state that:

> Retrofitted SUDS may prove useful in any situation where inadequate stormwater management leads to poor performance of the urban drainage system. This includes problems associated with excessive Combined Sewer Overflow (CSO) discharges, separate storm sewer outfalls and flooding of urban watercourses. Retrofitting also has a potentially important role to play in tackling sewer flooding risk; expanding the capacity of sewers and drainage systems to take on more 'load' deals with just one side of the supply-demand balance.
>
> EA (2007)

SUDS typically make use of a combination of permeable surfaces for source control; permeable conveying through swales and infiltration trenches; and shallow ponds with emergent vegetation, such as reed beds, for passive treatment over long retention times. A swale is a shallow ditch lined with vegetation, whilst an infiltration trench incorporates a layer of gravel. Ponds can also be constructed to attenuate storm water: referred to as balancing ponds or detention basins.

According to Environment Agency guidance on the design of SUDS:

> There are many SUDS design options to choose from and they can be tailored to fit all types of development, from hard surfaced areas to soft landscaped features. They can also be designed to improve amenity and biodiversity in developed areas. For instance, ponds can be designed as a local feature for recreational purposes and to provide valuable local wildlife habitat nodes and corridors.
>
> EA (2008)

In other words retrofitting SUDS can provide significant enhancement to ecological value. Even for a building such as a supermarket that has landscaping entirely devoted to car parking, permeable paving and other sustainable drainage features can be introduced. Porous surfaces and swales can be retrofitted to car parking areas, along with native plant species and trees that both increase ecological value and provide shading for parked cars (see Figures 12.4 and 12.5).

Green roofs and walls

Just as with landscaping, green roofs and walls can enhance the local ecology and contribute to local cooling effects. Green roofs are sometimes, perhaps more accurately, known as vegetated roofs. The principle is very simple, and has been around for centuries, i.e. turning a frequently neglected area of the building into a landscape. This can either be used as an amenity

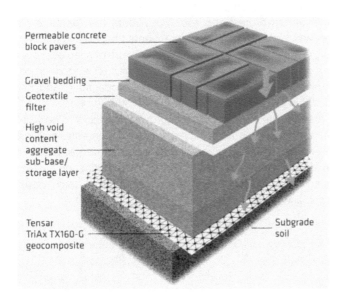

Permeable concrete block pavers

Gravel bedding

Geotextile filter

High void content aggregate sub-base/ storage layer

Tensar TriAx TX160-G geocomposite

Subgrade soil

Figure 12.4 Example of permeable block paving make up suitable for car park SUDS

Source: Tensar International Ltd

Figure 12.5 Swale under construction with completed one behind
Source: Duk

space, such as a roof garden, also known as an intensive green roof; or as a relatively lightweight covering providing a substrate for native plant species, or an extensive green roof.

Extensive green roofs sit on top of a waterproof membrane and comprise further layers to protect the roof from roots that penetrate a drainage layer, over which sits a filter mat that prevents the growing medium from washing into the growing layer. The most common plant species grown on extensive green roofs are those within the *Sedum* genus, or stonecrops. These flowering plants are low maintenance, self-propagating and are available in varieties that are native to most of the UK and Europe.

Modern green walls require a vertical growing medium to be fixed to the wall surface, planted with small, creeping herbaceous perennials, ferns, grasses and small shrubs. This can be used to create attractive patterns and textures. However, this system may require an integrated irrigation system to be built into its structure, normally employing hydroponics, which would add to the installation and maintenance cost.

MATERIALS

Embodied impacts
One of the key advantages of refurbishment over new build is the opportunity to reuse existing building materials. This may be in situ: for example, the

building shell; or through removing and reconditioning existing materials or components, such as material for crushing to hardcore, bricks or roof slates and in some cases windows.

According to the introduction to the UK Building Research Establishment's (BRE) *Envest* environmental assessment tool website:

> From material extraction, processing, component assembly, transport and construction, to maintenance and disposal, construction products have an environmental impact over their entire life cycle (such that):
>
> * 10 per cent of the UK CO_2 emissions arise from the production and use of building materials.
> * Each year the UK construction industry uses 6 tonnes of building materials per head of population.
> * Materials production and construction accounts for an estimated 122 million tonnes of waste, or 30 per cent of the total arising in the UK.[11]

Of course every component that is used in the refurbishment of a building has an embedded environmental impact. Each component will have gone through a process prior to construction, which may vary from adaptation or reconditioning of a reused component, through to a complex industrial process involving extraction of numerous raw materials, processing, assembly and delivery. Once installed, some components will need periodic maintenance or replacement using additional materials and processes. Eventually, as with new buildings, the whole assembly may need to be demolished and disposed of, refurbished again or adapted for another use.

The environmental impacts of importance that are used to determine Ecopoints for BRE's Envest scheme include:

* climate change – embodied carbon in kg CO_{2eq}, i.e. the greenhouse gas (GHG) emissions giving global warming potential (GWP) over 100 years;
* water extraction – mains, surface and groundwater extraction;
* mineral resource extraction – virgin irreplaceable materials such as ores and aggregates;
* stratospheric ozone depletion – chlorinated and brominated gases that destroy the ozone layer;
* human toxicity potential (HTP) – based on the full life cycle of the material;
* ecotoxicity to freshwater – maximum tolerable concentrations in water for ecosystems;
* ecotoxicity to land – maximum tolerable emissions to land;
* high-level nuclear waste – volume of high-level waste requiring long-term storage before it may be safe;

- waste disposal – tonnes of solid waste going to landfill or incineration in terms of loss of resource;
- fossil fuel depletion;
- eutrophication – over-enrichment of water courses causing algal growth and oxygen depletion;
- photochemical ozone creation – emissions of oxides of nitrogen (NO_x) and volatile organic compounds (VOCs) that convert to ozone in presence of sunlight;
- acidification – emissions of sulphur dioxide (SO_2) and NO_x that lead to acid deposition (acid rain).

Ecopoints form the basis for BRE's online Green Guide library,[12] and are used to establish materials credits within the Building Research Establishment Environmental Assessment Method (BREEAM) protocols.

A detailed account of the sustainable sourcing and procurement of materials for new buildings (equally applicable to retrofitting) is given in *Integrated Sustainable Design of Buildings* (Appleby, 2011), Chapter 3.12 of which deals with materials specification, including a review of the schemes available for assessing and rating the life cycle environmental impacts of construction materials. Chapter 4.2 sets out the principles of sustainable procurement, including modern methods of construction.

Responsible procurement

Purchasing of components for refurbishment should not only take into account these embodied impacts but also follow a process that BRE refers to as 'responsible sourcing'.

Sourcing of construction products occurs in at least two stages: products and materials will be specified to meet the cost and sustainability criteria to varying levels of detail depending on how critical they are to the refurbishment; followed during the refurbishment works by value engineering and procurement based on exact specifications and measured quantities.

For example, the BREEAM protocol is based on measurable standards of environmental management in the manufacture of materials from 'cradle to gate'. The protocol refers to four 'tiers' of quality, with Tier 1 meeting all of BRE's criteria. These require certification that certain standards of organisational and supply chain management and environmental and social responsibility have been met. There are a number of well-established national and international certification systems that do this, including the Forest Stewardship Council (FSC) and the Programme of Endorsement of Forest Certification (PEFC) schemes for timber.

Any responsible procurement strategy will include local sourcing of materials where feasible to reduce the energy used for transport.

Robustness

A final aspect of sustainable materials choice is life expectancy and replacement. Life cycle assessment of the environmental impact of building materials is based on an assumed life expectancy for each material. Mechanical and electrical plant and equipment may be subject to reduction in efficiency through use or breakdown due to mechanical failure. Materials on the other hand may be subject to weathering or wear and damage through attrition, abuse or moisture. Life expectancy for materials will therefore depend on the selection of suitably robust materials for the level of abuse expected.

Clearly the whole of the external envelope must be designed to exclude penetration of moisture into the building from precipitation, whilst prevention of surface and interstitial condensation requires suitable positioning of insulation and damp-proofing. Also robust construction depends on the use of materials that will withstand wear and tear. This is rewarded in BREEAM, for example, by incorporating a credit: 'to recognize and encourage adequate protection of exposed elements of the building and landscape, therefore minimizing the frequency of replacement and maximizing materials optimization'.[13]

During the planning of refurbishment it is useful to review the movements of people and vehicles around the building and carry out an 'impact' assessment on the routes they take around the building. This should identify the following:

- external locations where vehicle manoeuvres are required close to walls or soft landscaping that might need protection, such as loading bays and lawns;
- internal areas in which fork lift trucks, or similar will be manoeuvred;
- internal areas through which trolleys will be wheeled;
- internal areas with high pedestrian traffic loads;
- areas that are likely to experience above-average soiling and hence high cleaning intensities.

The following mitigation measures might be expected to increase the life expectancy of surfaces exposed to the above risks:

- bollards, barriers and/or raised kerbs protecting external walls in delivery and vehicle drop-off areas;
- robust external wall construction up to 2 metres from the ground; this could mean avoiding fragile rain screens, such as hung slates, at this level;
- corridor walls to be suitable for 'Severe Duty' (SD) as specified in BS 5234-2 (BSi, 1992);
- protection rails fitted at an appropriate height along corridor walls;
- protection to doors from impact by trolleys, etc.;

- hard-wearing and easily washable floor finishes in heavily used areas, such as main entrances, corridors, public areas, etc.

ENVIRONMENTAL ASSESSMENT RATING

Most environmental assessment schemes include credits for most of the aspects included in this chapter, as well as operational energy use. Environmental assessment schemes are described in Chapter 10 and are important for demonstrating the overall success of strategies to maximise environmental performance of retrofitting project.

NOTES

1. www.oecd.org/dataoecd/42/27/34416097.pdf (accessed 14 November 2013).
2. http://envirowise.wrap.org.uk/uk/Press-Office/Press-Releases/Wales/Spending-a-penny-could-be-costing-companies-in-pounds.html (accessed 14 November 2013).
3. http://wtl.defra.gov.uk/criteria.asp?technology=00030012&sub-technology=000300120001&partner=&section=2&submit_=Search&tech=000300 120001 (accessed 14 November 2013).
4. http://wtl.defra.gov.uk/criteria.asp?technology=00030014&sub-technology=000300140001&partner=&section=2&submit_=Search&tech=000300 140001 (accessed 14 November 2013).
5. www.allianceforwaterefficiency.org/commercial_dishwash_intro.aspx (accessed 14 November 2013).
6. http://sd.defra.gov.uk/advice/public/buying/products/electrical/dishwashers/standards/ (accessed 14 November 2013).
7. www.energystar.gov/index.cfm?c=comm_dishwashers.pr_crit_comm_dishwashers (accessed 14 November 2013).
8. www.ukrha.org/content.php?t=12&page=press (accessed 14 November 2013).
9. http://wtl.defra.gov.uk/technology.asp?technology=00030010&sub-technology= 000300100002&partner=&section=66&submit_=Search&tech=000300100002 (accessed 14 November 2013).
10. www.defra.gov.uk/environment/waste/ (accessed 14 November 2013).
11. http://envest2.bre.co.uk/detailsLCA.jsp (accessed 14 November 2013).
12. www.bre.co.uk/greenguide/podpage.jsp?id=2126 (accessed 14 November 2013).
13. www.breeam.org/filelibrary/Technical%20Manuals/SD5073_BREEAM_2011_ New_Construction_Technical_Guide_ISSUE_2_0.pdf (accessed 14 November 2013).

REFERENCES

Appleby, P. (2011) *Integrated Sustainable Design of Buildings*, London: Earthscan.
Appleby, P. (2013) *Sustainable Retrofit and Facilities Management*, Oxford: Routledge.
BSi (1992) *BS 5234-2 Partitions (Including Matching Linings). Specification for Performance Requirements for Strength and Robustness Including Methods of Test*, London: British Standards Institute.
CIBSE (2005) *Reclaimed Water* (CIBSE Knowledge Series), London: Institute of Energy.
EA (2006) *Managing Japanese Knotweed on Development Sites*, Bristol: Environment Agency. Available at: www.environment-agency.gov.uk/static/documents/Leisure/Knotweed_CoP.pdf (accessed 23 April 2012).

EA (2007) Science Report – SC060024, *Cost-Benefit of SUDS Retrofit in Urban Areas*, Bristol: Environment Agency. Available at: http://publications.environment-agency.gov.uk/PDF/SCHO0408BNXZ-E-E.pdf (accessed 24 April 2012).

EA (2008) *Sustainable Drainage Systems (SUDS)*, Bristol: Environment Agency.

EPA (2008) *Inventory of U.S. Greenhouse Gas Emissions and Sinks: 1990–2006*, Washington, DC: US Environmental Protection Agency. Available at: http://epa.gov/climatechange/emissions/downloads/08_CR.pdf (accessed 18 April 2012).

IEA (2010) *Key World Energy Statistics*, Paris: International Energy Agency.

13
ON-SITE CONSTRUCTION
Paul Appleby

No refurbishment project can meet sustainability objectives if they are not embedded into the construction process. Like quality, health and safety and value management, sustainability has to be bought into by all members of the supply chain. The 'Soft Landings' principles described in Chapter 14 facilitate continuity through the design, construction, commissioning, post-occupancy evaluation and aftercare processes.

This chapter examines the specific issues around organising and undertaking sustainable refurbishment for a range of project types and complexities. It is clear that refurbishment and retrofit can range from relatively minor works on an existing building through to near complete demolition and reconstruction, retaining perhaps some external walls or main structural elements, for example. These projects may be procured by building owners or property/facilities managers through to property developers. Indeed, major refurbishments may require an approach to construction and project management resembling new build in most respects. For a detailed analysis of sustainable construction see Part 4 of *Integrated Sustainable Design of Buildings* (Appleby, 2011). The chapter is based on practices and regulations in the UK to give insight into the wide range of issues related to sustainable refurbishment on the construction site. Practices and regulations vary greatly in different countries.

TENDERING FOR SUSTAINABLE REFURBISHMENT

Appointment of contractors that understand the challenges of sustainable refurbishment and have experience of the principles and technologies required is vital to the success of the project. In particular the framework established through the environmental assessment process commenced during the design stage must be maintained (see Chapter 10).There are numerous forms of contract for construction projects applied internationally, and most enable the incorporation of sustainability criteria, whilst some are better suited to refurbishment than others.

The key differentiator for refurbishment projects is the need for flexibility to allow for variations in the definition and scope of works. For example, prime cost contracts such as the UK developed JCT 2011: Prime Cost Building Contract, which is based around a design specification with an estimate of prime costs; and term contracts, such as the ICE C&C Term Version, which is also suitable for both planned and reactive maintenance. Measured term contracts involve the initiation of individual works by instructions as part of a programme of work, and priced according to rates related to the categories of work likely to form part of the programme.

The composition of the construction contract has a major impact on the success of translating the design team's sustainability objectives into practice. Some key decision points have been proposed by Joint Contracts Tribunal Ltd (JCT) as follows:

- The long and short term objectives you wish to achieve in terms of sustainability.
- What you must achieve in terms of sustainability i.e. legal and regulatory requirements, both project specific and general.
- Who is to take the lead in delivering the sustainability requirements. This will have a major influence on the procurement path to be followed e.g. management team, design team or construction team led.
- How to involve the supply chain in the design and construction process as early as practicable.
- Whether the providers of services and works are to be incentivised. If so, what mechanisms are to be used e.g. bonus, gain share, damages.
- What documentation is required e.g. specification, contract schedule, for the various providers and how the requirements are to be spelled out – are they to be generic provisions or specific measurable provisions.
- What items are to be provided for – consider the practicality of delivering sustainability requirements.
- Whether performance indicators are to be used, if so, which indicators and what target levels.
- The contract(s) to be used and which contract clause options related to sustainability are to be included. Insofar as it can, ensure the contract achieves certainty so that it is clear whether the requirements are fulfilled.

Joint Contracts Tribunal Ltd (2009)[1]

Achieving sustainability targets, along with control and management of quality, health and safety, whole-life costs and construction impacts in general, is greatly enhanced through collaboration of all parties. This is also a key element of the 'Soft Landings' approach. Although designed for use by employers who procure work on a regular basis and want to capture the benefits of long-term relationships within the supply chain, Framework Agreements are often used to provide a vehicle for partnering or collaborative approach.

Typically a JCT Framework Agreement will stipulate that:

The Provider will assist the Employer and the other Project Participants in exploring ways in which the environmental performance and sustainability of the Tasks might be improved and environmental impact reduced. For instance, the selection of products and materials and/or the adoption of construction/engineering techniques and processes which result in or involve:

- reductions in waste;
- reductions in energy consumption;
- reductions in mains water consumption;
- reductions in CO_2 emissions;
- reductions in materials from non-renewable sources;
- reductions in commercial vehicle movements;
- maintenance or optimization of biodiversity;
- maintenance or optimization of ecologically valuable habitat;
- improvements in whole-life performance.

It is usual for performance indicators to be incorporated into the contract document either as part of the specification or as a schedule to the contract conditions.

In this context the Building Research Establishment Environmental Assessment Method (BREEAM), or equivalent, assessment spreadsheet makes a very useful tracking tool for the client to monitor the performance of the construction team in achieving sustainability criteria. A good method of ensuring consistency and independence of this process is to use the same assessor to undertake periodic checks of the rating, reporting directly to the client, through to the final post-construction assessment.

Other performance indicators that are worth considering for inclusion in the schedule referred to above include Building Emission Rating (BER) as required for the Energy Performance Certificate and Building Regulations compliance in the UK (assuming the project requires Building Regulations approval), including any commitment for renewable energy generation as may have been required by the local planning authority, Site Waste Management Plans and relevant Constructing Excellence Key Performance Indicators (KPIs) (www.kpizone.com/).

There are several hundred KPIs, many of which address aspects of building and contractor performance that are not related to sustainability or refurbishment, hence it is important to relate relevant KPIs to the commitments made for the environmental assessment and development design, including socio-economic aspects, most of which will be location-specific. As described in Chapter 14, Soft Landings is not a target or KPI as such, but sets an approach for ensuring a continuity of approach from project inception through to post occupancy aftercare. Hence the contract must incorporate clauses that enable the contractors to integrate Soft Landings into their procedures, and price for them accordingly.[2]

MANAGING A REFURBISHMENT PROJECT

Managing sustainability requires translating design specifications into purchase and installation, whole-life cost control, control of construction impacts, management of 'considerate constructor' issues, waste management, value management, document control, commissioning, handover, post-construction assessment and aftercare. Most of these issues are mirrored in the management of health and safety on site and hence many of the control measures and responsibilities set out in such legislation as the UK's *Construction Design & Management (CDM) Regulations* can easily be adapted to sustainability and management of construction impacts.

As with CDM, all of these processes have to be trickled down through the supply chain through a combination of early involvement and incentivisation, established in the forms of contract referred to above. Refurbishment will usually incorporate some form of deconstruction or modification of existing building components up to and including demolition of fabric and structure. Whereas with new build projects any demolition may be either physically separate or programmed to provide a clear site prior to construction, refurbishment projects may involve part demolition within an occupied building, and have to be phased to minimise disruption.

The UK Institute of Demolition Engineers (IDE) provides brief guidelines on the management of demolition projects in accordance with the *Construction Design & Management (CDM) Regulations* 2007 (IDE, 2011). These are essentially a series of checklists that can be used as an *aide memoire*, the key issues from which are relevant to any refurbishment which involves intrusive work on fabric or structure as follows:

Asbestos

In the UK the management of asbestos is legislated through the *Control of Asbestos Regulations 2012* and the Health and Safety Executive (HSE) has produced associated guidance on the management, handling and survey of asbestos (see Figure 13.1).[3] Scrutiny of existing asbestos registers and details of previous removal and control measures is an important starting point, followed by a fully intrusive pre-demolition assessment survey, which is a statutory requirement under Regulation 10 of the CDM Regulations.

Figure 13.1 Typical enclosure for asbestos removal

Source: Wikimadia Commons

Guidance for these is provided in the HSE's downloadable publication: *HSG264: Asbestos - The Survey Guide*, Appendix 1 of which states that 'Refurbishment and demolition surveys are technically more challenging than management surveys, as their purpose is to identify all ACMs (asbestos-containing materials) within a particular building area or within the whole premises, so they can be removed' (HSE, 2012). This includes materials which may not have been visible to building managers but may be disturbed during refurbishment or demolition.

Other contaminants, including micro-organisms such as *Legionella*

Strategies should also be put into place to manage other sources of contamination to prevent or control emissions of oils, fuels, hazardous chemicals, airborne contaminants and micro-organisms. The processes involved will be very similar to that required for asbestos identification and management. Any hazardous substances or potential pollutants found in the existing building should have been identified and recorded as part of a CoSHH Assessment (*The Control of Substances Hazardous to Health Regulations 2002*), whilst any water systems that may present a risk of *legionellosis* should have been identified and maintenance records accessed.

The demolition or deconstruction risk assessment should identify these hazards and all necessary precautions instigated to minimize risk to operatives and those affected by their activities.

Utilities and existing services

Refurbishment may involve anything from minor alterations to existing services through to their complete removal and replacement. Any services that enter, leave or cross the site, including meters, substations, switchgear, stopcocks etc., should be identified and those that are to be retained provided with the necessary protection. Alterations that require notification to utility providers and/or changes to wayleaves or easements must be reflected in the programme and budget.

Party wall agreements

Any work on a 'party wall' that impacts on a neighbour comes under the auspices of the *Party Wall etc. Act 1996*. A party wall agreement or award would normally be agreed between neighbours providing the following protection under the Act:

For the building owner:

- there is less likely to be a dispute over responsibility should any damage occur to the adjoining owner's property;
- provides for a right of access under the Act to enter upon the adjoining owner's land to carry out work.

For the neighbour/adjoining owner:

- sets out the hours during which the work may be carried out;
- provides for making good any damage caused;
- ensures that the contractor has adequate public liability insurance in place.

Environmental impact: noise and vibration, air quality, contaminated land, ecology

In the UK not many refurbishment projects will require a formal environmental impact assessment (EIA) under the 2011 *Town & Country Planning (Environmental Impact Assessment) Regulations*; however, this does not mean that environmental impacts can be ignored. Projects that require planning approval may fall under Schedule 2 of the Regulations and be screened by the Local Authority to determine whether an EIA is required. Screening will also determine the scope of the EIA, including the impact of demolition, deconstruction, refurbishment and associated construction activities. For more details refer to Chapter 1.4 in Appleby (2011) and the wide range of guidance available on the Regulations such as those produced by the Institute of Environmental Management & Assessment (IEMA, 2004).

SUSTAINABLE SITE PRACTICES

Management of construction impacts

A building during refurbishment can be a hostile place, particularly if demolition is required, and its impact on the environment and its neighbours in particular is likely to be orders of magnitude greater than the building during normal operation. It is the responsibility of the 'principal contractor', usually the site manager, to minimize these impacts to an acceptable level. There are numerous activities carried out on any given site, from making tea to demolition, which vary enormously in their potential impact. Many of these activities are potentially disruptive and some can involve generating noise or pollution to land, watercourses or air. We have already seen that some of these are governed by statute and, in England and Wales, for example, some polluting activities will fall under the ambit of the 1990 *Environmental Protection Act* and require permitting under *The Pollution Prevention and Control Regulations 2000* with licences issued by the Environment Agency (Scottish Environmental Protection Agency – SEPA – in Scotland, which has its own Regulations). Others will fall under the 1974 *Health & Safety at Work etc. Act* and the CDM Regulations.

The emissions associated with construction activities are addressed in a number of ways. They are assessed under various headings in the EIA, if one is required for a planning application, whilst the carbon emissions associated with delivery and site activities form part of the life cycle assessment and carbon footprint of the development. Activities that can be controlled, monitored and recorded by the contractor are covered by the Construction Site Impacts credits in a BREEAM assessment. Some of these impacts are regulated and in some cases require licensing or permits.

The issues covered by BREEAM fall under the headings of: Monitoring, reporting and setting targets for:

- CO_2 emissions associated with site activities;
- CO_2 emissions associated with transport to and from site;
- water consumption associated with site activities.

Implementation of best practice policies to minimize:

- air (dust) pollution from site;
- water (ground and surface) pollution from site activities;
- use of sustainable site timber.

BREEAM also requires the contractor to have a policy for the sourcing and procurement of 'environmental' materials and an Environmental Management System in place.

Site waste management

Management of waste from a refurbishment project can present greater challenges in some respects than for a new build project. This is because of the potentially greater constraints on space and movements, particularly for a building or site that remains occupied during the refurbishment process. Handling demolition or deconstruction waste presents even greater challenges (see Figure 13.2).

Figure 13.2 Flexible chutes for conveying rubble from multi-storey building refurbishment

Source: HSS Hire

At more than 100 million tonnes per annum, construction and demolition waste represents approximately one-third of the total waste produced in the UK and about 25 per cent of the total mass of materials used in construction.[4] Of this around 30 million tonnes goes to landfill, whilst the government target for 2012 was to reduce this by 50 per cent.

In England since the introduction of the *Site Waste Management Plan Regulations* in 2008 there has been a requirement to produce a 'site waste management plan' (SWMP) for all construction projects with a projected cost of more than £300,000 on one site at the start of the project. An exception to this would be a project that falls within the ambit of *The Environmental Permitting (England & Wales) Regulations 2007*, which includes a requirement for waste management. In summary, the SWMP Regulations require the following as a minimum:

- The 'client' is responsible for preparation of the SWMP before construction starts and ensuring that its requirements are incorporated into contract documents for appointment of the 'principal contractor'.
- A 'principal contractor' must be identified who takes ownership of the SWMP and is responsible for keeping it up to date, trickling its requirements down to subcontractors, ensuring direct labour is trained in its requirement and complying with the waste management licensing, waste duty of care and waste carrier registration regimes. This is likely to be the same individual identified as 'principal contractor' in the *Construction (Design and Management) Regulations 2007*.
- The plan will need to describe a set of 'waste management estimates' against which performance will be compared as the refurbishment project progresses. First, the quantity of each type of waste likely to be produced on site will have to be forecast, along with the proportion that will be reused or recycled on site, or removed from the construction site for reuse, recycling, recovery or disposal elsewhere. As a minimum the waste will have to be classified as inert, non-hazardous or hazardous (as defined in the *Hazardous Waste Regulations 2005*).
- During construction the principal contractor must update the plan as waste is disposed of, reused or recycled. In this way the SWMP becomes a 'living' document that describes progress against the waste management forecasts contained in the plan.
- The duty of care requires the principal contractor to take care of waste while it is in their control, which includes:
  o checking that the person to whom the waste is given is authorised to receive it, making sure they hold the appropriate licence or permit;
  o completing, exchanging and keeping waste transfer notes when the waste is handed over and, for projects of

£500,000 and more, details of the destiny of waste once off-site;

o   taking all reasonable steps to prevent unauthorised handling or disposal by others.

- Performance against the SWMP should be reported following completion of the project. A more detailed analysis is required for projects of £500,000 or more, including an estimate of cost savings from site waste management.
- The SWMP must be kept for at least two years following completion of the project.

See Box 13.1 for a summary of the minimum information required in a SWMP.

BOX 13.1 MINIMUM CONTENTS OF A SITE WASTE MANAGEMENT PLAN

Responsibilities:
1   the client;
2   the principal contractor;
3   the person who drafted the plan.

Description of the Construction Works:
4   the location of the construction site;
5   the estimated cost of the project.

Materials Resource Efficiency:
6   any decision taken before the SWMP was drafted to minimize the quantity of waste produced on site.

Waste Management:
7   description of each waste type expected to be produced during the project;
8   for each waste type estimate of the quantity of waste that will be produced;
9   for each waste type the waste management action proposed (including reuse, recycling, other types of recovery and disposal).

Waste Controls and Handling:
10   a declaration that all waste produced on the site will be dealt with in accordance with the waste duty of care;
11   a declaration that materials will be handled efficiently and waste managed appropriately.

Recycling demolition material

Approximately 275 million tonnes of aggregates are used each year in the UK as raw construction materials, with around 70 million tonnes derived from recycled or secondary sources. The quality of the recycled aggregate is dependent upon the quality of the materials that are processed, the selection and separation processing used, and the degree of final processing that these materials undergo. Depending on the volume of material generated, significant benefits are possible with on-site sorting and crushing, particularly if a proportion can be stored and reused on site. These benefits include reduced transport costs and environmental impacts. If it is necessary to purchase or hire crushing plant and/or lease a nearby site to locate it then these benefits may be eroded considerably. Crushing demolition material is a noisy process and the plant takes up considerable space, so it may be difficult to arrange for on-site aggregate recycling where there are major space restrictions, as may be the case for a typical refurbishment project.

Considerate constructors

In the UK, the Considerate Constructors Scheme (CCS) was set up in 1997[5] following other initiatives and studies to improve the image of the construction industry. BREEAM rewards compliance with this scheme which is applicable for any site in the UK. CCS is reviewed annually and it is important to use the most recent checklist when establishing what actions and resources are required to achieve the targeted score under the Scheme.[6] CCS is a voluntary scheme that encourages the considerate management of construction sites. The scheme is operated by the Construction Confederation and points are awarded under the following headings:

- enhancing appearance;
- protecting the environment;
- securing everyone's safety;
- caring for the workforce.

In addition information is gleaned on any innovative practices and feedback from the site manager, as well as site specific data relating to the workforce, accidents and environmental incidents.

To achieve 'Certification of Compliance' under this scheme a score of at least 5 is required under each heading, including some mandatory requirements. 'Certification beyond Compliance' is awarded if a score of 7 or more is achieved under each heading. BREEAM awards one credit if a score of 24–34 is achieved, with a minimum of five points under each heading, and two credits for a score of 35–9, with at least 7 under each heading. In addition BREEAM requires a score of at least 40 as a prerequisite for an overall BREEAM Exemplary Rating.

The Construction Federation rewards the top 7.5 per cent of schemes every year with bronze, silver or gold awards. The scheme uses trained independent 'monitors' to review the performance of a site against the checklist referred to above and based typically on two visits. Each issue needs to be validated by reference to clear and verifiable records.

MONITORING SUSTAINABLE PRACTICES ON SITE

The CCS, or its equivalent, can also be used to provide a vehicle for continuous monitoring of sustainability performance as well as assessing overall performance. In concert with this, whether or not a BREEAM Assessment is being carried out, the protocol for BREEAM 2011 New Construction (see Chapter 11) (which also applies to major refurbishment) provides a useful tool for monitoring construction impacts.

Compliance requires that:

- Responsibility has been assigned to an individual(s) for monitoring, recording and reporting energy, water and transport consumption data resulting from all construction processes. To ensure the robust collection of information, this individual(s) has the appropriate authority, responsibility and access to the data required.
- Data on energy consumption (kWh) from the use of construction plant, equipment (mobile and fixed) and site accommodation necessary for completion of all construction processes should be monitored and recorded.
- Using the collated data the energy consumption (total kWh and kWh per £100k of project value) and greenhouse gas emissions (as total kg CO_2 equivalent and kg CO_2 equivalent per £100k of project value) from the construction process should be reported. In the case of a BREEAM Assessment this requires the use of a bespoke scoring and reporting tool.
- Data on potable water consumption (m^3) from the use of construction plant, equipment (mobile and fixed) and site accommodation necessary for completion of all construction processes should be monitored and recorded.
- Using the collated data the total net water consumption (m^3), i.e. consumption minus any recycled water use, from the construction process should be reported (e.g. via the BREEAM scoring and reporting tool).
- Data on transport movements resulting from delivery of the majority of construction materials to site and construction waste from site should be reported. As a minimum this should cover:
  o transport of materials from the factory gate to the building site, including any transport, intermediate storage and

distribution. Scope of this monitoring must cover the following as a minimum:

- materials used in major building elements, including insulation materials;
- ground works and landscaping materials.

    o   transport of construction waste from the construction gate to waste disposal processing/recovery centre gate. Scope of this monitoring should cover the construction waste groups outlined in the project's SWMP.

- Using the collated data, separate reports can be prepared for materials and waste, indicating the total fuel consumption (litres) and total greenhouse gas emissions (kg CO_2 equivalent), plus total distance travelled (km). A tool for this is provided within the BREEAM scheme.

INTEGRATION WITH COMMISSIONING PROCESS

Although commissioning is often considered to be an activity that occurs once construction activity has finished, good practice is for the process to be fully integrated into the construction process. In the USA it is common practice to appoint a 'commissioning authority', who reports directly to the client, prior to tendering the refurbishment contract. Indeed this is a pre-requisite of a Leadership in Energy and Environmental Design (LEED) assessment, whilst BREEAM rewards the appointment of a project team member to monitor and programme pre-commissioning, commissioning and, where necessary, recommissioning on behalf of the client. For more complex installations a 'commissioning manager' should be appointed at the design stage in order to undertake reviews of 'commissionability' and provide input to programming and installation, as well as managing the whole commissioning process through to handover and aftercare in line with the Soft Landings principles. See Chapter 14 for a more detailed review of Soft Landings and the commissioning, handover and aftercare processes.

HANDOVER AND AFTER

Clearly the whole design, refurbishment and commissioning process is leading to the handover of a building to the client which is fully operational and performing as the designers intended. However, handover should not become a cliff-edge event since in all but the simplest of refurbishments there may need to be ongoing seasonal commissioning, post-occupancy evaluation and aftercare service as set out in the Soft Landings Framework.

The UK Building Services Research and Information Association (BSRIA) provides guidance on the application of the Soft Landings principles in meeting the requirements of a BREEAM assessment:

> The Soft Landings Framework encourages a programme of follow-through, with fine-tuning and seasonal re-commissioning. The Framework distinguishes between an initial period of aftercare (the first eight weeks) and extended aftercare which may last up to three years.
>
> BSRIA (2011)

For refurbishment projects that follow the Soft Landings principles both the client and the design and refurbishment teams should have appointed Soft Landings champions. Logically the client-side champion could be drawn from the facilities management team. The supply-side champion should be appointed at project inception and if possible novated to the refurbishment contractor (including demolition where applicable). Along with the commissioning manager/authority these champions provide the necessary coordination and continuity to ensure that sustainability measures drawn up during the design process are percolated throughout the entire construction, commissioning and aftercare process. As well as championing Soft Landings principles these same individuals can be responsible for coordinating both BREEAM assessment tasks and CCS monitoring.

NOTES

1 This reference, although still available online at the time of writing was revised in 2011, although this issue is not available as a free download.
2 Online article by Roderic Bunn of BSRIA can be found at: www.bsria.co.uk/news/soft-landings-budgets/ (accessed 25 April 2013).
3 Most of these are free to download via the HSE's publications website at www.hse.gov.uk/asbestos/index.htm (accessed 4 April 2013).
4 From UK Office for National Statistics at www.statistics.gov.uk/cci/nugget.asp?id=1304 (accessed 5 April 2013).
5 www.ccscheme.org.uk/ (accessed 7 April 2013).
6 The most recent Site Registration Monitors' Checklist can be downloaded from: www.ccscheme.org.uk/index.php/site-registration/site-managers-information/registered-site-checklist (accessed 7 April 2013).

REFERENCES

Appleby, P. (2011) *Integrated Sustainable Design of Buildings*, London: Earthscan.
BSRIA (2011) *BG 28/2011: BREEAM 2011 & Soft Landings – An Interpretation Note for Clients and Designers*, Bracknell: Building Services Research and Information Association. Available at: www.bsria.co.uk/bookshop/books/breeam-2011-soft-landings/ (accessed 24 April 2013).

HSE (2012) *HSG264: Asbestos – The Survey Guide*, Health and Safety Executive. Available at: www.hse.gov.uk/pubns/books/hsg264.htm (accessed 7 April 2013).

IDE (2011) *Management of Demolition Projects*, Chatham: Institute of Demolition Engineers. Available at: www.ide.org.uk/pdf/guidelines_for_clients&admins_2011-03-07.pdf (accessed 6 April 2013).

IEMA (2004) *Guidelines for Environmental Impact Assessment*, Lincoln: Institute of Environmental Management & Assessment.

Joint Contracts Tribunal Ltd (2009) *Guidance Note: Building a Sustainable Future Together*, London: Sweet & Maxwell. Available for purchase at: www.constructionbooks.net/uploaded_files/1131/images/guidance_note_-_building_a_sustainable_future_together.pdf

The Construction (Design and Management) Regulations 2007. Available at: www.opsi.gov.uk/si/si2007/uksi_20070320_en_1 (accessed 30 March 2010).

The Control of Asbestos Regulations 2012. Available at: www.legislation.gov.uk/uksi/2012/632/pdfs/uksi_20120632_en.pdf (accessed 12 April 2013).

The Control of Substances Hazardous to Health Regulations 2002. Available at: www.legislation.gov.uk/uksi/2002/2677/contents/made (accessed 12 April 2013).

The Environmental Permitting (England & Wales) Regulations 2007. Available at: www.opsi.gov.uk/si/si2007/uksi_20073538_en_1 (accessed 30 March 2010).

The Hazardous Waste Regulations 2005. Available at: www.opsi.gov.uk/si/si2005/20050894.htm (accessed 30 March 2010)

The Pollution Prevention and Control (England and Wales) Regulations 2000. Available at: www.legislation.gov.uk/uksi/2000/1973/contents/made (accessed 17 April 2013).

The Site Waste Management Plans Regulations 2008. Available at: www.opsi.gov.uk/si/si2008/uksi_20080314_en_1 (accessed 30 March 2010).

The Town & Country Planning (Environmental Impact Assessment) Regulations 2011. Available at: www.legislation.gov.uk/uksi/2011/1824/made (accessed 23 April 2013).

14

INTEGRATING DESIGN AND USE: THE 'SOFT LANDINGS' PHILOSOPHY

Simon Burton
Source: Roderic Bunn, BSRIA

The Introduction highlighted the dangers of not taking sufficient account of user needs and expectations in the design process of refurbishment projects and the dangers of poor services set up and building management during use. Many buildings are known to suffer poor user satisfaction and energy performance after refurbishment and this clearly makes them unsustainable. Sustainable refurbishment must involve not only sustainable design and materials choice, etc. as described in previous chapters, but also ecological, social and financial sustainability in building operation for the life of the building. This chapter describes one comprehensive system which is being developed and used in the UK to ensure that new and refurbished buildings really are sustainable after handover.

The design team on a refurbishment project will receive briefing from several parties including the clients, owners and agents. Sometimes this will include the future occupants; however, the end users may not be known at the design stage, particularly with speculative developments. In addition, life does not stand still and particularly if the refurbishment process is lengthy, user requirements may well change during the building procurement. However well the design team fulfils its brief, the performance of the occupied building may not live up to the original intentions and calculations, leading to dissatisfaction and complaints. The design team may be blamed for this, but many factors can coincide to create a mismatch between expectations and operational outcomes. What is ideally needed is integration of user behavioural factors in the briefing and design process, a series of reality checks as procurement and on-site construction proceeds, followed by a period of professional aftercare including post-occupancy evaluation.

Figure 14.1 shows the various ways in which energy actually used in the completed building may vary from that anticipated in the design and modelling stage. Whilst additional working hours and additional equipment installed and used are acceptable reasons for increased consumption, other reasons such as imperfect management and control are bad news.[1]

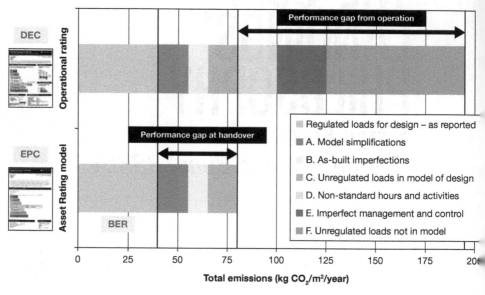

Figure 14.1 The energy performance gap
Source: Robert Cohen, Verco

WHAT IS 'SOFT LANDINGS'?

Soft Landings[2] helps clients and occupiers to get the best out of their new or altered buildings. It is designed to reduce the tensions and frustrations that so often occur during initial occupancy, and which can easily leave residual problems that persist indefinitely. At its core is a greater involvement of designers and constructors with building users and operators before, during and after handover of building work, with an emphasis on improving operational readiness and performance in use.

It is not just a handover protocol but also provides the 'golden thread' which links between:

- the procurement process – setting and maintaining client and design aspirations that are both ambitious and realistic, and managing them through the whole procurement process and into use;
- initial occupation, providing support, detecting problems and undertaking fine-tuning;
- longer-term monitoring, review, POE and feedback – drawing important activities into the design and construction process which are both rare in themselves and often disconnected.

Soft Landings extends the duties of the building team before handover, in the weeks immediately after handover, for the first year of occupation, and for the second and third years. In order to improve the chances of success, it reinforces activities during the earlier stages of briefing, design and construction. It creates opportunities for greater interaction and understanding between the supply side of the industry and clients, building users and facilities managers. It helps everybody concerned to improve their processes and products, and to focus innovations on things that really make a difference.

HOW IT WORKS

Soft Landings is not intended to be external consultant led, the client is an active participant and leads the process at the outset to develop the roles and responsibilities. This should include client representatives, all key design professionals and the supply chain. The people involved in this process should be the actual individuals who will work on the project.

Soft Landings can be used for refurbishment and alteration as well as for new construction. It is designed to smooth the transition into use and to address problems that Post-Occupancy Evaluations (POE) show to be widespread. It is not just about better commissioning and fine-tuning, though for many buildings commissioning can only be completed properly once the building has encountered the full range of weather and operating conditions. Soft Landings starts by raising awareness of performance in use in the early stages of briefing and feasibility, helps to set realistic targets and assigns responsibilities for the client and design team. It manages the realisation of expectations through design, construction and commissioning, and into initial operation, with particular attention to detail in the weeks immediately before and after handover (see Figure 14.2).

A period of professional aftercare, with monitoring, performance reviews and feedback aims to help occupants make better use of their buildings, while clients, designers, builders and managers gain a better understanding of what to do next time. It provides support throughout the whole procurement process, especially:

- During inception and briefing, to establish client and design targets which are better informed by performance outcomes in use on previous projects. It also commits those joining the design and building team to follow through after handover and for project management to begin to allocate responsibilities for ongoing reviews of design intent and anticipated performance, and to prepare for the other activities required.
- Alongside the design and construction process, to review performance expectations as the client's requirements, design solutions, and management and user needs become more concrete and the inevitable changes are made. An important element here is the carrying out of 'reality checks' to ensure that changes do not conflict with expectations nor have knock-on effects in other areas. In addition the team must plan for commissioning, handover and aftercare, and involve the occupier much more closely in decisions that affect operation and management.
- In the weeks before and after handover; although practical completion is important legally and contractually, with Soft Landings, handover is no longer the end of the job, but just an event in the middle of a more extended 'finish' stage. Before handover, the team prepares to deliver the building and its systems in a better state of operational readiness. When the occupants begin to move in, the aftercare team (or team member) will have a designated workplace in the building and be available at known times to explain design intent, answer questions, and to undertake or organise any necessary trouble-shooting and fine-tuning. Both before and after handover, the design and building team will work closely with the client, occupiers, and facilities managers to share experiences and smooth the transition into use.
- During the first three years of occupancy, to monitor performance, to help to deal with any problems and queries, to incorporate independent post-occupancy surveys (such as occupant satisfaction, technical and energy performance), and to discuss, act upon and learn from the outcomes. Achievements and lessons should then be carried back to inform the industry and its clients.

COSTS AND BUDGETING

The costs associated with using the approach will be relatively small for large projects and major refurbishments but may be more significant for small projects and minor refurbishments. This may be an impediment despite the

RIBA 2008 Stages	RIBA 2013 Stages	CIC stage and title 2012	Soft Landings	BSRIA BG6 – Design Framework stages	ACE Services (2009 Agreements)
	0 – Strategic Definition	0 – Strategic Definition			
A Appraisal	1 – Preparation and Brief	1 – Preparation and Brief		Stage 1: Preparation	Performance / Performance design in buildings (ACE schedule of services Part G(c)) / Performance plus (not a recognised ACE term but in common usage to extend duties) / Detailed design in buildings (ACE schedule of services Part G(b)) – co-ordinated working drawings are listed under other services to be specified separately
B Design brief			Stage 1. Briefing: Identify all actions needed to support the procurement		
C Concept	2 – Concept Design	2 – Concept Design	Stage 2. Design development: to support the design as it evolves	Stage 2: Design	
D Design Development	3 – Developed Design	3 – Developed Design	Pitstop 1: Scheme design reality check		
E Technical Design	4 – Technical Design	4 – Technical Design	Pitstop 2: Technical reality check		
F1 Production information				Stage 3: Pre-construction	
F2	Information Exchanges will vary depending on the selected procurement route and building contract. A bespoke RIBA Plan of Work 20§3 will set out the specific tendering and procurement activities that will occur at each stage in relation to the chosen procurement route				
G Tender documentation			Optional Pitstop revisit		
H Tender action			Pitstop 3: Tender award stage reality check		Mobilisation, construction and completion – builders' work details are under other services to be specified separately
J Mobilisation					
K Construction to practical completion	5 – Construction	5 – Fabrication Design		Stage 4: Construction	
	6 – Handover and close-out	6 – As constructed	Pitstop 4: Pre-handover reality check	Stage 5: Commissioning	
			Stage 3. Pre handover: Prepare for building readiness. Provide technical guidance	Stage 6: Pre-handover	
			Pitstop 5: Post-handover sign-off review. Ensure all outstanding Pitstop items are complete and system is signed off as operational	Stage 7: Initial occupation	
L1	7 – In Use	7 – In Use	State 4. Aftercare in the initial period: Support in the first few weeks of occupation		
L2 Post-practical completion			Stage 5. Years 1 to 3 Aftercare: Monitoring review, fine-tuning and feedback	Stage 8: Post-occupancy aftercare	
L3					

Figure 14.2 Alignment of the Soft Landings stages with the 2013 RIBA Plan of Work, and the 2007 edition of the CIC Scope of Services

Source: BSRIA

undeniable long-term financial benefits of better energy performance and occupant satisfaction and productivity. The costs can usefully be divided into those for the briefing, design and construction stages, often minimal compared with the conventional management, and those for the one or two years of post-construction monitoring and tuning, which could be seen as part of the maintenance and facilities management activities and thus the responsibility of the occupant.

Early experience with Soft Landings procurement suggests that there are a variety of ways in which clients are funding Soft Landings activities. Some clients are calling upon their professional designers and the main contractor to individually price out Soft Landings services. Others are including Soft Landings as one line in their employer's requirements and expecting Soft Landings activities to be described and provided by the supply chain with no input from the client. Some clients are setting aside budgets for the additional Soft Landings activities, while others are expecting something for nothing.

In instances where the client has not provided any detailed guidance on their requirements or budget availability, contractors and designers will be forced to itemise and price out their Soft Landings activities stage-by-stage in their tender returns. This may raise costs way above what the client was expecting, and may also lead to unnecessary duplication of roles and activities by different members of the bidding team.

Not only is this an inappropriate way to develop a Soft Landings process, but also bidders to the project will be motivated to price out their activities at a full commercial labour rate. This means the client will miss out on opportunities for discounting certain activities, particularly where the firms involved may stand to benefit professionally or commercially. All parties to the aftercare process stand to gain from the lessons learned process. Design consultants, particularly, could choose to be flexible on their costs.

That said, consultants normally only attend site after handover to help resolve defects. It may not be appropriate to link their help with resolving defects with their Soft Landings fine-tuning responsibilities. This is an uncomfortable mix and consultants will expect extra fees for additional professional services. Overall, it is counterproductive for the supply chain to be forced to cost out their Soft Landings activities in the absence of a strong steer from the client. Clients are therefore advised to take a more active role in budgeting and procuring Soft Landings services.

At the outset, clients should work with their professional advisors to consider which Soft Landings activities are needed (bearing in mind the essential requirements laid out in the Soft Landings Core Principles, see below), and work out what a reasonable budget might be. This process will help the client decide what will be included in the project budget and what can be funded from elsewhere. For example, certain elements of Soft Landings – such as the post-occupancy evaluations – might be best funded out of operational expenditure budgets rather than the project budget. Similarly, pre-design feedback analysis of a client's existing buildings could be funded separately from the project budget. (Even if the client requests this work from its professional designers, arguably it could be something that designers should undertake routinely to inform their design deliberations.)

When all these activities have been taken into account, the client can arrive at a budget that will inform which Soft Landings activities are to be defined and detailed in the project documentation. This will provide much-needed clarity for prospective tenderers.

Nevertheless, budgeting for Soft Landings may still be an approximate process. Some clients may prefer to create a provisional sum against which tenderers price their roles and activities. This could be the basis of a Soft Landings resource schedule. Bear in mind that notional budgets and provisional sums have a habit of being lost, or spent on other things. It may also be difficult to price out activities (with a reasonable degree of accuracy) for the period of extended aftercare three or four years hence. In this situation it would be acceptable for the first year period of aftercare to be fully budgeted, and a provisional sum set aside for the second and third years of aftercare, and/or a tariff card agreed for aftercare activities.

If the client can ensure that the aftercare budget is listed in the bill of quantities as something that must be ring-fenced by the contractor, a provisional sum is an appropriate way of protecting the aftercare funds. Clients still need to be explicit about the activities covered in the aftercare contract so that there is no uncertainty about what is and what is not included, such as the post-occupancy evaluations. This should prevent any arguments later, and any accidental double-pricing by the contractor.

OTHER BENEFITS RESULTING FROM SOFT LANDINGS

Other important but less directly tangible benefits for the architect and project manager include: client retention owing to the improved levels of service; greater mutual understanding between designers, builders, clients, occupiers and managers; and education of design and project team members in what works well and what may be causing difficulties. In the longer term it also helps to develop industry skills in problem diagnosis and treatment.

BRINGING THE WHOLE PROJECT TOGETHER

Many techniques of project feedback and POE are aimed at one particular stage of a project or to suit a single discipline or element such as building services engineering. Many are used solely in the post-occupation phase when it is too late to tackle the strategic problems that originated in briefing, design and project management. Soft Landings provides a process carrier for these techniques, so helping to unite all disciplines and all stakeholders and to extend the procurement process beyond handover. As POE becomes more routine, findings and benchmarks from previous POE surveys can be used to help calibrate client and design expectations. Where practicable, results from these surveys can also provide metrics that allow these expectations to be tracked from briefing, through design development, construction and commissioning, and into operation.

THE FIVE STAGES

The procedures are designed to augment standard professional scopes of service, not to replace them. They can be tailored to run alongside most industry standard procurement routes potentially in any country, to create the most appropriate service to suit the project concerned. Major revisions to industry standard documentation are not necessary. The main additions to normal scopes of service occur during five main stages:

1   Inception and briefing to clarify the duties of members of the client, design and building teams during critical stages, and help set and manage expectations for performance in use.

2   Design development and review (including specification and construction). This proceeds much as usual, but with greater attention to applying the procedures established in the briefing stage, reviewing the likely performance against the original expectations and achieving specific outcomes.

3   Pre-handover, with greater involvement of designers, builders, operators and commissioning and controls specialists, in order to strengthen the operational readiness of the building.

4   Initial aftercare during the users' settling-in period, with a resident representative or team on site to help pass on knowledge, respond to queries, and react to problems.

5   Aftercare in years 1–3 after handover, with periodic monitoring and review of building performance.

The Soft Landings Framework is an open-source procedure published by BSRIA. A detailed description of the activities of the five stages and example worksheets are available from the Soft Landings website.[3]

THE TWELVE PRINCIPLES

Soft Landings is an integrated suite of activities, but even so clients and professional users may find that they only have the time and budget to include some components. In this situation the following twelve principles are suggested as essential:

1   Adopt the entire process: The project should be procured as a 'Soft Landings' project, and project documentation should explicitly state that the project team will adopt the five work stages of the Soft Landings Framework to the greatest extent possible.

2   Provide leadership: The client should show leadership, engender an atmosphere of trust and respect, support open and honest collaboration, and procure a design and construction process that can be conducted with equal levels of commitment from all disciplines.

3   Set roles and responsibilities: In Soft Landings, the client is an active participant, and leads the process at the outset to develop the roles and responsibilities. This should include client representatives, all key design professionals and the supply chain. The people involved in this process should be the actual individuals who will work on the project.

4   Ensure continuity: Soft Landings should be continuous throughout the contractual process. It should be made part of all later appointments, and expressed clearly in contracts and subcontract work packages as appropriate. The client and main contractor should ensure that subcontractors and specialist contractors take their Soft Landings roles and responsibilities seriously.

5   Commit to aftercare: There should be a clear and expressed commitment by the client and project team to follow through with Soft Landings aftercare activities, and to observe, fine-tune and review performance for three years post-completion. The aftercare activities should aim to achieve the Soft Landings performance objectives, and any targets agreed at the design stage.

6   Share risk and responsibility: The client and main contractor should create a culture of shared risk and responsibility. Incentives should be used to encourage the project team to deliver a high-performance building that matches the design intentions.

7   Use feedback to inform design: The client's requirements, the design brief and the design response should be informed by performance feedback from earlier projects. The desired operational outcomes need to be expressed clearly and realistically.

8   Focus on operational outcomes: The Soft Landings team should focus on the building's performance in use. Regular reality checking should be carried out to ensure that the detailed design and its execution continues to match the client's requirements, the design team's ambitions and any specific project objectives.

9 Involve the building managers: The organisation that will manage the finished building should have a meaningful input to the client's requirements and the formulation of the brief.

10 Involve the end users: Prospective occupants should be actively researched to understand their needs and expectations, which should inform the client's requirements and the design brief. There should be a clear process for managing expectations throughout the construction process and into building operation.

11 Set performance objectives: Performance objectives for the building should be set at the outset. They should be well-researched, appropriate and realistic, capable of being monitored and reality checked throughout design and construction, and measurable post-completion in line with the client's key performance indicators.

12 Communicate and inform: Regardless of their legal and contractual obligations to one another, project team members need to be comfortable communicating with the entire team in order to achieve the levels of collaboration necessary to carry out Soft Landings activities.

SUMMARY

The Soft Landings approach is an example method of how refurbishment projects can be helped to be really sustainable in their actual operation, not just in their design intentions. It can be summarised in five basic points:

- It provides a unified vehicle for engaging with outcomes throughout the process of briefing, design and delivery. It dovetails with energy performance certification, building logbooks, green leases and corporate social responsibility.
- It can run alongside any procurement process. It helps design and building teams to appreciate how buildings are used, managed and maintained.
- It provides the best opportunity for producing low-carbon buildings that meet their design targets. It includes fine-tuning in the early days of occupation and provides a natural route for post-occupancy evaluation.
- The additional costs are well within the margin of competitive bids. During design and construction, Soft Landings helps performance-related activities to be carried out more systematically. There is some extra work during the three-year aftercare period, but the costs are modest in relation to the value added to the client's building.

- Soft Landings creates virtuous circles for all and offers a good method for achieving truly integrated, robust and sustainable design.

NOTES

1   *Closing the Performance Gap between the Predicted and Actual Energy Performance of Non-Domestic Buildings.* Paper prepared by Robert Cohen for the Green Construction Board Buildings Working Group. May 2012.
2   The Soft Landings Framework, BSRIA BG 4/2009, June 2009.
3   www.softlandings.org.uk (accessed 14 November 2013).

15
CASE STUDIES

The case studies have been chosen to reflect a range of building types, uses, refurbishment levels, locations, methodologies and solutions. The buildings chosen are 'commercial' both in terms of their use – offices, retail and industrial buildings – and their ownership. However, two public office buildings are included (Elizabeth II Court and the NVE Building) as they demonstrate important comprehensive solutions incorporating most of the issues described in this book, and increasingly the differences between public and private, ownership and use, is becoming more blurred. Public buildings often act as test beds and lead the way forward for the private sector. Other owners are developers and owner-occupiers, many of whom have some sort of vested interest in promoting sustainable refurbishment, either as it is important for their particular business or for their public image. Large and small buildings (Kreditanstalt für Wiederaufbau to Cullinan Studios) are included, though it seems that the issues and solutions adopted can be similar.

TECHNICAL ASPECTS

The systems and technologies chosen are clearly affected by what remains of the building – the floor-to-floor heights, the need to maintain listed architectural elements, noise for adjacent roads, etc. and this may have restricted available approaches. However, many innovative systems and technologies were used, together with the more conventional energy-saving measures commonly seen in new commercial buildings.

It is interesting that 'Passive House' standards for insulation, airtightness and energy use have been adopted in two of the case studies (ebök Vermögensverwaltung Office and AS Solar Headquarters) and insulation to high standards was installed in all buildings. Only two of the studies used only natural ventilation (Cullinan Studios and Scotstoun House), though several others used natural ventilation as a part of their ventilation strategy and included opening windows. Reducing solar gain with shading, mostly external shading where this was possible, was seen as important in most studies. Night ventilation to cool the structure of the building was used in at least five of the case studies (Elizabeth II Court, Printing Office Karlsruhe, Scotstoun House, NVE Offices and ebök Vermögensverwaltung Office), and is likely to be used in practice in others. To enhance the effects, phase-change materials (PCM) were built into three of these buildings (Printing Office Karlsruhe, ebök Vermögensverwaltung Office and Scotstoun House). Absorption chillers were installed in two examples (Kreditanstalt für Wiederaufbau and AS Solar Headquarters) one using solar heat and the other heat from the district heating system. Space heating used district heating networks in the NVE Offices and the Centro Colombo Shopping, with waste heat used in three other examples (Elizabeth II Court, Printing Office Karlsruhe and AS Solar Headquarters). An air-sourced heat pump only seems to have been used in one study (Cullinan Studios) but biomass heating was used in three studies (Angel Building, AS Solar Headquarters and Scotstoun House). Heat recovery was installed on all mechanical ventilation systems except one which had a large supply of waste heat. Maximising the use of daylighting and sophisticated control of artificial lighting is common to all of the examples. Renewable energy sources (photovoltaics, solar thermal and/or bio-mass boilers) were retrofitted in four examples (Cullinan Studios, Angel Building, AS Solar Headquarters and Scotstoun House), with one (AS Solar Headquarters) becoming an energy-positive building. In addition, ground sourcing was installed in two projects (Printing Office Karlsruhe and ebök Vermögensverwaltung Office). Occupant comfort and adaptive opportunities for occupants are built into several projects reflecting the interest in occupant satisfaction and presumably related productivity. Most of the case studies also included several of the other environmental issues as described in Chapter 12 and achieved high environmental rating assessments.

Most of the case studies are of buildings needing radical retrofitting and upgrading so that they have been vacated and stripped back, sometimes to the basic frame. This means that the systems and measures adopted are little different from those used in new build projects; Building Regulations are similar to new build and the old building becomes virtually a new building. This is usually claimed to be more 'sustainable' than demolition and total rebuilding, due to the embodied energy in the frame and foundations, and the energy used in demolition and disposal of the rubble. While this is likely to be true, most of the case studies were chosen as they demonstrate how more sustainability features have been built into the renovated building, in most cases going beyond those required by Building Regulations.

Retrofitting of historic listed buildings naturally gives rise to more complexity and this is reflected in four of the case studies (Cullinan Studios, NVE Building, ebök Vermögensverwaltung Office and Scotstoun House). All managed radical conversion to modern uses with high energy standards.

Only one of the case studies is about upgrading a building with occupation continuing (Centro Colombo Shopping) but this does not reflect what is going on in most countries. Renovation with occupants in place or decanted within a building complex is common but tends to be unreported. It can greatly increase sustainability and lead to very efficient buildings (Baker, 2009: Chapters 11 and 12).

COUNTRY DIFFERENCES

Different countries always vary in their approach to treating their old building stock. The USA has a much higher demolition and rebuilding rate than other countries, some countries like the UK have many listed buildings that must not be demolished but retained and maintained for posterity, while Germany seems to take a very comprehensive approach to retrofitting its existing building stock. The case studies presented here come from four countries demonstrating that there is interest and activity widely in sustainable refurbishment of commercial buildings. Although the approaches and emphasis vary between countries, the issues addressed and the solutions used are similar and variations seem to be due more to the buildings and locations than to different country views of what 'sustainability' should encompass. However, even if there is consensus on the issues, the case studies demonstrate a wide range of methods of implementation.

ACHIEVEMENTS AND MONITORING

It is unusual to find a refurbishment project that has dealt well with every aspect of sustainability and has demonstrated best performance in energy use, other environmental issues, occupant comfort and satisfaction, financial viability and replicability. There are many obvious reasons for this. Thus, in the case studies presented here, focus is on the best aspects of each project, and most have many, to demonstrate what can and has been achieved. Thus, on looking through all the studies, the reader can see the total picture of what is achievable and hopefully see how future projects can achieve five-star results.

One of the most elusive aspects of 'sustainable' building projects is the assessment of post-construction performance in terms of both energy use and occupant satisfaction, as discussed in Chapter 14. This has many causes, including the length of time after handover needed for the building to settle down and be 'tuned' to optimise performance (a minimum of one year), the need to monitor energy consumption for at least a year after stability has been achieved and adjust for weather conditions, and the need to carry out and analyse adequate occupant surveys. The basic questions to be answered are: Does the building perform in use as intended by the designers? Is the result a truly sustainable building? Part of the process must include whether a change in use or occupancy of the building from the design brief is the cause of different performance or whether design, construction, equipment or management aspects are responsible. Performance monitoring is included in the case studies where it is available (Elizabeth II Court, Karlsruhe Printing Office and ebök Vermögensverwaltung Office) but is sadly lacking in many cases where sustainable refurbishment is claimed. In all the three monitored projects, forecast energy and comfort standards were nearly achieved and further lessons were learnt about how the standards might be fully achieved, though this did require further work in some areas. Readers should bear in mind that good design intentions and implementation do not make a sustainable refurbishment; the proof of sustainability is in the monitored performance.

CASE STUDY 1: ELIZABETH II COURT, HAMPSHIRE, UK

Comprehensive retrofitting of 1960s offices to high energy standards

- remodelled to fit with the local environment;
- ventilation chimneys added;
- opening windows and night cooling;
- mixed-mode ventilation;
- total monitored energy use 131 kWh/m²/year, 7 per cent higher than original design;
- Building Research Establishment Environmental Assessment Scheme (BREEAM) 'Excellent'.

Elizabeth II Court is Hampshire County Council's former Ashburton Court site in the centre of Winchester, a historic town in the south of England. The building comprises North, West and East blocks on a podium car park. Hampshire County Council decided to refurbish the entire site and turn it into an exemplar of sustainable and energy-efficient office space.

The three-storey East block, built in the 1960s, is typical of local authority offices of the period: heavyweight concrete construction with prefabricated concrete panels and single-pane glazing with horizontally pivoting openable windows for natural ventilation.

Ashburton Court suffered from an aggressive external presence which jarred with the sensitive historic context of Winchester and had attracted national notoriety. The refurbished design articulates the building in a more sensitive manner through reduced massing, new elevations and use of local materials. A new scale, rhythm and modulation has been introduced which softens the presence of the building and relates to the rich architecture of the city.

Client: Hampshire County Council
Architect: Bennetts Associates
Quantity Surveyor: Davis Langdon
Structural Engineer: Gifford
Services Engineer: Ernest Griffiths
Main Contractor: BAM

Figure 15.1 The transformed building sits more comfortably with the scale, rhythm and materiality of Winchester

Source: © Tim Crocker

Figure 15.2 Original building

Source: Bennetts Associates

Refurbishment

The condition and thermal performance of the fabric was poor by contemporary standards with overall facade U-values varying between 1.6 to 5 W/m²K. The building was far from airtight, lighting was also very typical: fluorescent luminaires fitted in a suspended ceiling. This ceiling not only hid the building services, but also disconnected the building's thermal mass from the occupied space. The lighting control zones were too large for any energy-saving measures to be effective.

In 2006, a Building Use Studies (BUS) occupant satisfaction survey of the East block revealed extremely unhappy occupants. The building was ranked near the bottom of the BUS dataset, one of the lowest scores in twenty years of surveys. People were highly uncomfortable in summer and winter, there was too much electric lighting and not enough natural light, and the offices were too noisy.

The refurbishment programme was phased to reduce the cost of decanting staff during construction. Staff continued to work in the North and West blocks while the East block was refurbished during Phase 1, the North and West blocks being refurbished over the following eighteen months. The redeveloped 3,000 m² East block was completed in December 2008.

The design team were appointed on the basis of the low-energy refurbishment proposed in their bid.

The architect and the mechanical and electrical consultant had worked together before, and were familiar with blending architecture and engineering. With committed players, the design team were also able to deal with the challenges thrown up by the project such as cost, programme, logistics and value engineering. The project's scale, complexity and phased construction meant the early involvement of a major contractor to ensure that a collaborative approach was taken to the project's development and implementation.

The form of the building, with long narrow floor plates and tight floor-to-ceiling heights, lent itself to a low-energy refurbishment using natural ventilation rather than a four-pipe fan coil system. The design team wanted to establish an environmental strategy that was sufficiently robust and flexible to accommodate change with an emphasis placed on sound environmental engineering.

The existing building was stripped back to its structural concrete frame. Then in order to allow wind-driven ventilation, ducts were placed on

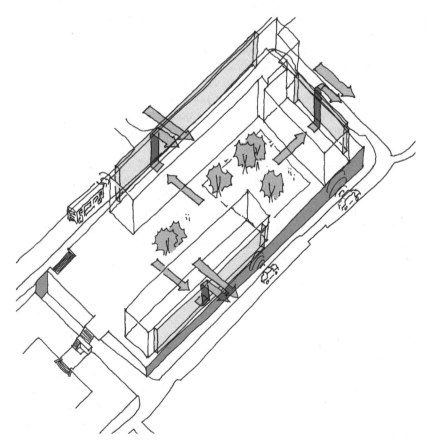

Figure 15.3 Ventilation strategy
Source: Bennetts Associates

the outside facades. Ventilation air is drawn into the building from the court-yards, across the floor plate and then up acoustically attenuated ventilation ducts on the street facades of the building. The 'wind troughs' on top of the ducts exploit wind blowing across the roof to create suction, which draws air through the building. The brick-clad ventilation ducts also helped to shade the east and west facades and to bring a more appropriate scale and rhythm to a building that had been very horizontal and disconnected from the street.

All the buildings on the site were extensively remodelled. The top floors of the West and East blocks were cut back to reduce the overall massing and to reduce the perceived height of the building. The car park space at podium level under the East block was changed to office accom-modation, with the similar area under the West and North blocks used for a new IT suite and ancillary spaces. This included a new reception, a cafe and auditorium pavilions. These changes saw the accommodation increased from around 9,000 m^2 to nearly 14,000 m^2, of which the East block represents about a quarter.

There are two main types of facade, the west-facing street elevation of windows and brick-clad ventilation chimneys, and the courtyard-facing elevation of windows and aluminium panels.

Elizabeth II Court's elevations face east and west, which presented a challenge in terms of solar control. The facade design has attempted a high degree of self-shading. Glazing was selected to optimise solar control and light transmission properties. The large number of openable windows meant that air leakage performance needed to be good. Internal blinds (for solar and glare control) also needed to take account of natural air paths, so perforated blinds were used.

The floor-to-floor heights on the office floors vary around 3.35 m, as determined by the dimensions of the original structure. The removal of the suspended ceiling exposed the original concrete slab of 880 mm coffers, increased the effective floor-to-ceiling height and released the thermal capacity of the building to assist in moderating internal temperatures. The installation of a raised floor meant that the effective floor-to-ceiling heights remained virtually unchanged.

Energy refurbishment

Energy modelling

Early energy calculations suggested fuel consumption of 59 kWh/m^2/year for fossil fuel and 34 kWh/m^2/year for electricity. The targets were subsequently refined to 57 kWh/m^2/year for fossil fuel and 66 kWh/m^2/year for electricity. The combined CO_2 emission target of 39 kg CO_2/m^2/year was equivalent to a 10 per cent improvement over the *Energy Consumption Guide 19* (*ECON 19*, 2003) good practice performance for a hybrid office, with a 15 per cent Type 3 component. The CO_2 emission target was subsequently reduced to 35 kg CO_2/m^2/year as more information became available during design development. Modelling was carried out using TAS and additionally wind tunnel testing at Cardiff University.

Solar controls and shading

The main facades face east and west, which are both difficult to shade due to low-angle morning or afternoon sun and required vertical rather than horizontal solar shading. The majority of solar shading is provided by the depth of the ventilation ducts, with additional vertical louvre blades.

Figure 15.4 Shading devices on the west facade

Source: © Tim Crocker

Ventilation and cooling

Traffic noise was the central driver to the chosen ventilation strategy. The top floor of the East block is well above street level and facades are set back. This enables it to be cross-ventilated on both sides of the building. The wind-driven cross-ventilation makes good use of the pressure differentials developed across the top of the building.

On the lower floors, the street elevation is equipped with ventilation chimneys, located at regular intervals along the facade. Each extract chimney serves one of the three lower floors. An offset at the top of each chimney prevents rain penetration, improves the acoustic properties and provides a location for a vertical motorised opening sash. The window bays on the courtyard facade have a Building Management System (BMS)-controlled motorised top light that can be opened to give a free area of 50 per cent. The lower windows can be opened manually to a free area of 30 per cent. The motorised windows and the chimney sashes are automatically controlled to vary the amount of natural ventilation. Local manual overrides are provided for each group of automatic windows and chimney sashes with BMS override to limit the period of manual operation.

The windows and chimney sashes are automatically opened in summer whenever room temperature exceeds 16°C at day or night or when CO_2 levels reach certain levels. They are also opened at night by the BMS for night purging when external temperatures exceed set levels, taking advantage of the thermal mass of the exposed concrete soffits.

Openable windows have been provided to give occupants the freedom to control their ventilation needs, enabling them to trade off air movement, temperature, air quality and outside noise, as they see fit.

The mixed-mode mechanical ventilation system is used to deliver minimum outdoor air ventilation during colder weather and also during hot weather. A small supply air-handling unit is provided at each level, adjacent to each core. Supply fans are variable speed controlled on CO_2 readings. Supply air is filtered, heated if necessary and ducted into the floor void.

Heating

Much of the ventilation heating load is satisfied by waste heat from the data centre, and this is only topped up by the boilers when external conditions dictate. It was decided that in these circumstances there was no need for exhaust heat recovery and the associated fan pressure drops.

The office heating system is served by low-temperature hot water, provided by three gas-fired condensing boilers. Under normal operating conditions two boilers satisfy demand. Each partitioned room has a small fan under the floor, which draws air from the adjacent open-plan floor void and supplies the room via floor diffusers. Extract air is exhausted naturally via the chimneys and via copier and toilet extract systems. Open office areas are heated by perimeter radiators fitted with thermostatic radiator valves. Partitioned rooms are heated by heat recovery variable air volume (VAV) units, controlled by wall-mounted controllers.

Lighting

Lighting in the open-plan office areas relies on suspended fluorescent fittings with an element of up-lighting to illuminate the exposed waffle slab. The luminaires are arranged in rows across the width of the building at about 2.6 m intervals to coincide with the spacing of the waffles. Each section of light fittings has a passive infrared movement detector and a programmable light level sensor. The lamps are dimmed down to off when an area is unoccupied for more than a predetermined, programmable period. There are no manual switches, nor is there a central control system.

Other environmental refurbishment

Embodied energy

With low-energy buildings, embodied energy can account for as much as 20 per cent of the life cycle environmental impact. This project reuses the existing in-situ concrete frame. The foundations and structural frame account for approximately half the embodied energy of a building.

Noise control

The building is surrounded on three sides by heavily trafficked roads. This prevented a reliance on ventilation using opening windows on the street facades. The ventilation ducts and wind troughs allow wind-driven cross-ventilation without needing to open street-facing windows.

Water

It was not possible to locate any water recycling tanks under the building so all efforts were made to reduce water consumption by installing efficient fittings. Despite increasing the number of people on the site it was important to ensure that no more drainage was generated, due to the capacity of the surrounding drainage system.

Materials

The most significant materials issue was keeping the sub- and super-structures. In addition, the bricks were sourced local, partly for aesthetic reasons but also partly driven by transport energy. The precast concrete cladding panels removed from the old building were crushed and used as hardcore in other Hampshire County Council projects.

Energy reduction and environmental achievements

An environmental rating of BREEAM 'Excellent' was achieved. Hampshire County Council prepared a thirty-page booklet *Ashburton Court East – A Guide for Staff*. The guide included space plans, types of workspace and office equipment. One page covered the heating, ventilation and lighting systems.

Figure 15.5 Transformed modern office interior and diversity of space supports flexible working

Source: © Tim Crocker

The East block was completed in December 2008 and performance monitoring was carried out of this block during a twelve-month period, from November 2009 to October 2010. It is important to stress that energy measurements in the first eighteen months will not be representative of a building's long-term performance. There are many variables: outstanding defects, delayed commissioning, phased occupation and fine-tuning of systems to suit occupants' needs.

The twelve-month monitored results gave an overall energy use of 131 kWh/m^2/year, 7 per cent higher than the original design target. Table 15.1 shows how the energy consumption breaks down by end use.

At 77 kWh/m^2/year, the total electrical consumption of the East block was around 17 per cent higher than the original design target. For lighting, the average monthly consumption was around 5,700 kWh, or 21.9 kWh/m^2/year. Including an adjustment factor of 1.4 kWh/m^2/year for the lighting in the stairs and stair lobbies, the corrected annual consumption was 23.3 kWh/m^2/year. Domestic hot water electrical consumption was 4.6 kWh/m^2/year (close to the good practice benchmark), cooling close to 10 kWh/m^2/year, and fans and pumps 5.6 kWh/m^2/year, with a proportion of this caused by pumping for heat reclaim from the IT suite.

With gas consumption at 54 kWh/m^2/year, the East block achieved slightly better than the original design target. This excludes heat reclaim equivalent to around 17 kWh/m^2/year. It has been suggested that savings of 11–21 kWh/m^2/year could be made, for example, by reducing heating during summer and by fine-tuning the interaction between heating and natural ventilation throughout the year. With these measures gas consumption could be brought down to 43–33 kWh/m^2/year.

Table **15.1** Energy consumption breakdown

	Elizabeth II Court (East block) energy consumption in kWh/m2 per annum (3185 m2 TFA)								
	Econ 19 Type 3 Good practice	Econ 19 Type 2 Good practice	Original building metered	Original building modelled	Business as usual model	Design Target	LCBA Phase 2 model	LCBA Phase 3 model	*Metered data
Heating and hot water	97	79	187	300	56		28	39	54
Hot water		1					1	1	5
Cooling	14	4	0	0	12		4	7	10
Fans and pumps	30	22	12	45	28		3	4	6
Lighting	27	22	31	27	25		26	24	24
Office equipment	23	20	42	15	30		33	29	28
Other electricity	7	4	1	0	2		1	1	6
Total gas	97	79	187	300	56	57	28	39	54
Total electricity	101	51	86	87	97	66	68	66	77
Total	198	130	273	387	153	123	96	105	131
	Elizabeth II Court carbon dioxide emissions in kgCO2/m2 per annum								
Gas	19	15	36	58	11	11	5	8	10
Electricity	43	22	36	37	41	28	29	28	33
**Total	61	37	73	95	52	39	34	35	43

Source: Bennetts Associates

*Monitoring from November 2009 to October 2010
**At carbon factors of 0.194 for gas and 0.422 for electricity

Financial aspects

The construction cost associated with the project was £29.17M or £2,317/m² at a base date of November 2005. The total project value including professional fees, furniture, fittings and equipment, car park repair works, limpet asbestos removal, decanting and temporary accommodation costs, was £40.17M at a base date of November 2005.

Getting the building into operation

Commissioning the building was carried out in January 2008. A second BUS occupant survey was carried out in July 2010. The twelve summary variables showed a huge improvement in the occupants' view of the refurbished building compared with the pre-refurbishment survey. It scored particularly well in terms of design, occupant needs and image.

In summer, the control of temperature was generally acceptable, in the range 22–26°C for most of the time during working hours. Control of the peak operational temperature (the comfort temperature experienced by occupants) could be improved by optimising night cooling. Control of ventilation was generally good both in summer and in winter, although a balance needs to be struck between indoor air quality, occupant comfort and heating energy consumption. Apart from maintaining closed windows during weekends when night cooling is possible, the automatic windows were generally operating as intended: fully open during warmer spells, modulating during cooler periods, and closed when the external temperature is below 15°C.

Although perceived productivity scored average, it represented a step change improvement over the original building. Overall, the East block scored in the top 40 per cent of the UK benchmark data set. Performance might have been even better had there not been a difficult eighteen months of initial occupation as the other blocks of Elizabeth II Court were being completed.

Temperature control has been generally effective both in winter and summer, although there is some scope for optimising night cooling. Control of ventilation using the CO_2 level as an indicator has been generally effective both in both summer and winter.

However, teething problems with the CO_2 sensors caused draughts. According to the building manager: 'We found that the CO_2 sensors were the controlling sensors in the winter, so we got to a point where the chimneys opened and cool air was dumping on people sitting close to them. We raised the carbon dioxide set point slightly, and that seemed to resolve the issues without any problems of "stuffiness"'.

The automatic windows have operated well, although it was noticed that early morning sun tends to penetrate deep into the building, causing occupants to drop blinds because of glare. Many do not lift them back up later, which leads to 'blinds down and lights on' syndrome.

The data centre heat reclaim system appears to be meeting the mechanical ventilation heat demand in all but the coldest months.

The facilities contractor reported that the building works well from an operational perspective but is quite complex. Dealing with staff complaints is therefore something of an issue. The response to complaints about lighting and control tends to be reactive and therefore rather ad hoc. The most challenging day is one that is warm and windy, as some staff complain about draughts and feeling cold. Internal blinds also flutter in the wind. Every floor tends to be different but the higher up the building the windier it gets. Hot, still days tend to be better. The spring and autumn are the trickiest.

In the three years since the building was occupied, occupant numbers have increased to 270–80 persons, up from the design density of 250.

Final conclusion

The East block of Elizabeth II Court was a classic opportunity for demonstrating low-carbon refurbishment. It had a committed client and suitably experienced and enthusiastic design team. The frame of the original building was ideal for maximising opportunities for natural ventilation and daylighting.

Figure 15.6 Central courtyard after refurbishment

Source: © Tim Crocker

CASE STUDY 2: PRINTING OFFICE, KARLSRUHE, GERMANY

Comprehensive office refurbishment focusing on energy and comfort, monitored for five years

- new highly insulated facade;
- heating from waste heat, ground cooling, night cooling, use of PCM;
- 58 per cent overall energy saving;
- comfort greatly increased;
- conclusion – some poor mechanical design and poor construction blamed for initial poor performance before optimisation.

This office building, owned by E&B Engelhardt und Bauer GmbH in Karlsruhe, was a typical commercial building constructed in 1978. It was refurbished in 2004 with the ambitious aim of the building owner and the building designers to develop an integral and sustainable building with an energy concept which features highly efficient and novel technologies for heating and cooling, with significantly reduced primary energy use, whilst respecting the occupants' requirements on thermal comfort.

The building exhibited the common weak points of this type of building: high energy demand (both electricity and fossil fuels), poor insulation, insufficient daylighting, unsatisfactory air quality, poor room acoustics and thermal discomfort in both winter and summer. The single-storey building had a gross area of 480 m^2 and a gross air volume of 1,250 m^3 and was densely equipped with office devices, giving measured internal heat gains of around 400 W/m^2.

Owner: E&B Engalhardt & Bauer, Karlsruhe, Germany

Partner: Karlsruher Institut für Technologie (fbta)

Monitoring: Fraunhofer-Institut für Solare Energiesysteme ISE

Figure 15.7 The original printing offices
Source: © Fraunhofer ISE

Refurbishment drive

In the course of the refurbishment, the building was enlarged by adding a second floor, increasing the gross area of 1,110 m² to a gross volume of 3,900 m³.

The general refurbishment concept included:

- A new building envelope using a lightweight steel construction.
- A higher insulation standard to reduce heat loss and the discomfort caused by both warm and cold surfaces and draughts.
- The removal of the suspended ceiling exposing the concrete ceiling, altering the previous space into a higher and brighter workplace. Moreover, it provided additional thermal storage for attenuation of the room temperatures and enabled the use of low-energy cooling in spite of high solar and internal heat gains.
- A new large glass facade opening the building to the west and providing a good view for all employees working in the office. The automated solar shading system reduces the solar heat gains and guarantees glare-free daylight conditions using a cut-off control algorithm.
- Improving the room acoustics in the offices by the new room geometry, textile partitions and the cooling panels.

re 15.8 Post-refurbishment offices

rce: fbta, Karlsruhe Institute of Technology

Energy efficiency design

The reduced heating and cooling loads of the refurbished building enabled the use of low-output heating and cooling systems. The large heat transfer area of the 'thermo-active building systems' (TABS) allowed effective operation with small temperature differences between the surface temperature of the TABS and the required room temperature. Water supply temperatures of 16–20°C in cooling mode and 26–30°C in heating mode, enabled the use of low-temperature heat sources and sinks.

Insulation

The new external structure was insulated to achieve U-values for the exterior wall of 0.3 W/m²K, for the base plate 0.27 W/m²K, for the roof 0.2 W/m²K and for the windows (glass + frame) 1.4 W/m²K. Taking into account thermal bridging, the overall U-value of the building was 0.54 W/m²K.

Cooling concept

In summer, the ground is used as an environmental heat sink directly to cool the building via the TABS. A field of twelve borehole heat exchangers with a depth of 44 metres provides 10 kW of cooling power. The ground floor is cooled by a capillary tube grid mounted on the solid intermediate ceiling. The upper floor uses novel radiant cooling panels with Micronal® PCM, suspended from the ceiling. In addition, night ventilation is used driven by the buoyancy effect using the night air to cool the building structure.

The control system takes the different thermodynamic characteristics into account; the thermally heavier PCM-radiant panels are operated during the night, while the thermally lighter grid conditioning system cools the space during the day.

Heating concept

The original heating strategy was that the entire building would be heated primarily by waste heat from the printing plant employing low-temperature radiators and radiant floor heating. Additionally, the existing gas boiler was to be available as back-up heating.

Ventilation concept

During occupancy, the ventilation strategy uses hybrid ventilation in the building depending on time and user behaviour. The mechanical ventilation works in conjunction with natural ventilation since the occupants can individually open the windows. The mechanical ventilation system operates mainly in winter when the building is occupied, depending on indoor and outdoor temperatures, and in summer when there are high outdoor temperatures. There is a heat recovery system which improves the air quality while reducing the heating load.

Figure 15.9 Installing the ground-sourced cooling

Source: © Fraunhofer ISE

The outline planning of the project was carried out by a combined team of the architect, the structural engineer, the heating, ventilation and air-conditioning (HVAC) planner, the cooling panel company, the borehole drilling company and a company for the shading systems. The building was finally refurbished by a general contractor.

Performance monitoring

After completion of the refurbishment works and reoccupation of the building, extensive monitoring of the building and plant was carried out for 2007 and 2008. Data were recorded by the building management system. The monitoring data consist of minute-by-minute measurements of temperatures as well as of flow, energy and power meters. Raw data were processed before data evaluation using a sophisticated method to remove erroneous values and outliers from the database. The results and conclusions of the monitoring study included the useful heating and cooling energy, auxiliary energy use for the pumping and the ventilation system, as well as the end and primary energy use, occupant thermal comfort and local meteorological site conditions.

The two-year monitoring campaign and the comprehensive data evaluation yielded important lessons, which led to improvements of the HVAC system and consequently to the overall building performance.

Performance problems uncovered

Despite thorough design planning of the building and the energy concept, the printing office building did not initially perform in practice as anticipated during the design stage. There were many reasons for this, including incorrect sizing of the systems, improper equipment selection and installation errors. The main reason, however, is believed to be due to the general contractor who did not show strong commitment to the energy-efficiency concept for the refurbishment.

After the refurbishment, the building still required a relatively high heating energy demand of 80 kWh/m² in 2008. The primary reason was that the building envelope was not constructed to Passive House standard as planned during the design stage. Second, the installed heat recovery unit of the ventilation system did not perform with the design efficiency of 85 per cent, thus requiring the retrofit of a supplementary heating system in order to preheat the incoming outdoor air.

Figure 15.10 Monitoring was initially carried out two years

Source: fbta, Karlsruhe Institute of Technology

The entire building was supposed to be heated primarily (around 90 per cent) by waste heat with radiant floor heating on the first floor and low-temperature radiators on the second floor. However, the waste heat from the printing plant contributed only 10 per cent (8 kWh/m²) to the total heating demand due to an incorrect installation and operation of the heat exchanger between the printing plant and heating system. The remaining heat load was supplied by the gas boiler. Moreover, the electrical energy used for lighting was unexpectedly high at 23.6 kWh/m^2, due to the workplaces having higher lighting requirements than planned.

The ground-sink-based cooling system was undersized due to an incorrect assumption in the simulation. An unusually high undisturbed ground temperature of 14–15°C was measured during the original Thermal Response Test (TRT). However, the size of the heat sink, i.e. the number and depth of the borehole heat exchangers, was calculated with a ground temperature level of 10.9°C, which is a typical value for the region of south Germany. As a result, the cooling output was only 10 kWh/m^2, half the calculated cooling demand to provide good thermal comfort to Class II standard.

In addition, the monitoring of the circuits of the cooling system revealed a high volume flow and high pressure drops due to the small dimensions of the pipe system.

Changes made

The whole TABS system was retrofitted and optimised in 2008 and a reversible heat pump of 33 kW output was installed. In winter, the heat pump now operates in combination with the gas boiler to provide heat to the building and cooling to the ground-sink system, while in summer the heat pump in conjunction with the ground sink provides the 20kWh/m^2 required to cool the building.

End and primary energy performance

The subsequent monitoring and the evaluation of the building and plant performance confirm that both the energy efficiency and the interior comfort were greatly improved. After optimisation of the services in the building, particularly the cooling system, the total primary energy use for heating, cooling, ventilation and lighting was reduced by 58 per cent of the figure before the retrofit, to 134 kWh/m^2 (2010 figure). The heating energy use itself was reduced by 52 per cent to 64 kWh/m^2.

Thermal comfort

The post-occupancy evaluation in 2008 shows high satisfaction with the overall building concept but the need for optimisation regarding thermal comfort. The occupants were satisfied with the indoor air quality, the light conditions in the workplace and the relative humidity. However, they perceived the room temperature as cool in winter and warm in summer. Due to the open-plan office, the occupants criticised the lack of individual possibilities to change the indoor environment. However, occupants felt more comfortable in the retrofitted building ('Neubau') than in the old part of the building ('Altbau').

In addition, thermal comfort assessments of the building in summer and winter were made according to the European guideline EN 15251:2007-

08, based on the number of hours during occupancy when the operative room temperatures are outside defined comfort classes.

Set points for lower limits of occupant thermal comfort in winter, independent of the ambient air temperature, are 21°C (class I), 20°C (class II) and 19°C (class III). During the winter period, operative room temperatures stayed above 21°C, thus indicating comfort class I.

Of the two approaches in EN 15251:2007-08 for comfort in summer, static and adaptive, it was decided to carry out the evaluation based on the adaptive model. The summer period is defined by a running mean ambient air temperature greater than 15°C. Comfort rating is measured by the number of hours exceeding a set temperature. Interior thermal comfort conditions during the monitoring years 2007 and 2008 satisfied the comfort class III. During higher ambient air temperatures (running mean above 20°C), the measured operative room temperatures exceeded 28°C and, for a few hours, even exceeded 30°C on the second floor. The cooling concept for the building aimed at a thermal comfort of class II, which was thus not achieved. According to the results for the summer season of 2008, the interior thermal conditions satisfied class I during 87 per cent of the working time and class II and class III during 5 per cent of the working time. The operative room temperature exceeded all three defined comfort boundaries for thirty-two hours.

Other environmental issues

The printing workshop is accredited according to EMAS, the EU Eco-Management and Audit Scheme.

Financial aspects

The full implementation costs for the renovation works were for the physical construction 860 €/m² and for the services systems 370 €/m².

Summary and conclusion

A sustainable refurbishment should aim at a high workplace comfort, a significantly reduced heating and cooling demand, highly efficient services and the use of renewable energy sources where possible. The printing office building in Karlsruhe is a good example that energy-efficient and cost-effective building concepts are not contradictory to high workplace quality. The energy concept used is capable of adaptation to other refurbishment projects, since the original building was a typical example of older office buildings.

However, only an integrated design approach to building and HVAC will result in a lean energy result, and provide high workplace quality. This project shows that the interaction between designers, building owner, general contractor and subcontractors can be a great difficulty for the achievement of energy-efficient building refurbishments. A well-organised tender and contract, and support of an energy 'mediator' to coordinate all energy-related issues is considered essential for the successful implementation of sophisticated projects.

Information provided by:
EnOB: Research for energy-optimised construction

Figure 15.11 Post-refurbishment space
Source: fbta, Karlsruhe Institute of Technology

'Buildings of the Future' is the guiding concept behind EnOB – research for energy-optimised construction (the name EnOB is an abbreviation of the equivalent German term Energieoptimiertes Bauen). The research projects sponsored by the German Federal Ministry of Economics and Technology involve buildings that have minimal primary energy requirements and high occupant comfort, with moderate investment costs and significantly reduced operating costs.

For more information see: www.enob.info/en/ (accessed 3 November 2013).

CASE STUDY 3: THE FOUNDRY, CULLINAN STUDIO, LONDON, UK

Nineteenth-century listed warehouse converted to offices, complete renovation retaining the main structural elements

- 70 kWh/m² for all energy use (calculated);
- high insulation and natural ventilation;
- heat pump with underfloor heating;
- BREEAM 'Excellent';
- PV provides 10 per cent of energy demand.

Edward Cullinan Architects was set up as a cooperative in 1965 and in November 2012 changed its name to Cullinan Studio. The practice is based in north London and has recently moved into its newly refurbished offices on the Regents Canal. In the 1990s recession, the practice bought Baldwin Terrace, a row of three disused industrial buildings along the canal, and when the lease on their existing offices was coming to an end they decided to refurbish part of the buildings as their own permanent offices. The original

Client and architect: Cullinan Studio

Mechanical and Electrical engineers: Max Fordham LLP

Contractor: Jerram Falkus Construction Ltd

building was a nineteenth-century Victorian warehouse and foundry using the canal for transport. The 800 m^2 building was a narrow three-storey brick block between the canal and the parallel service road, facing north/south with a slate pitched roof. The building had been occupied for many years on a temporary basis by small studios for artists and similar activities and was in fairly poor condition before it was decided to carry out the major refurbishment.

ure 15.12 The listed south facade wall
ng the canal

rce: Tim Soar

re 15.13 The first-floor space below the listed oak trusses

rce: Tim Soar

Refurbishment drive

Some aspects of the building were locally listed, the south facade along the canal and the top-floor roof trusses, and so these had to be retained and renovated in the refurbishment. The retrofit planning permission was granted in 2008 along with a residential development of twelve new flats in place of the old design studios next door in the terrace. Many iterations of the design were necessary to gain planning permission and subsequently to accommodate changing economic circumstances to allow the project to go ahead.

When the existing tenants had moved out, the building was cleared and rebuilding design work started.

Building surveys showed the existing south wall to be leaning up to 300 mm and several renovation options were explored including demolition of the whole building except the canal wall, storing the roof trusses off-site and rebuilding, or renovation of the building as it stood. The latter strategy was eventually chosen to minimise environmental impact.

The first and also decisive moves of the retrofit intervention were a response to two constraints of the existing foundry building – the leaning and locally listed south wall that needed stabilising and its double-height windows. Steel shear walls were chosen to articulate two circulation zones framing the studio spaces. A horizontal truss, with intermediate supports, spans between the shear walls, supporting the failing wall and creating a void against it to express the double-height windows of the listed wall. These moves not only stabilised the building but maximised views and light and combined to create a radical rearrangement of the existing building volume. The windows in the listed south wall were enlarged as much as was permitted by conservation guidance to maximise potential for daylight and ventilation.

The first floor has excellent natural light and cross-ventilation created by cutting the existing north wall off at first-floor sill level and providing a continuous band of glazing hung from the steel frame, both to stabilise the existing fabric and to support the listed timber trusses.

Refurbishment work started in 2011 and was completed in September 2012.

Energy refurbishment

Fabric insulation

The low-energy strategy was a fabric-first approach. The north and south walls achieve U-values of 0.1 W/m²K, using a variable thickness of 'Warmcell' (recycled newspaper) blown onto the inner face of the south wall between the outward sloping external wall and the vertical inner wall. The existing north wall uses externally applied expanded polystyrene insulation to a thickness of 380 mm over the existing rendered facade giving a U-value of 0.1 W/m²k. The internal floor was insulated with Rockwool, and the roof with Rockwool and Cellotex under a reclaimed Welsh slate roof finish. Draught-proofing and

1. Existing Fabric Retained
2. Recycled Newspaper
3. Natural Ventilation
4. Air Source Heat Pump
5. 22m² PV Panels
6. Heat Recovery
7. Underfloor Heating

ure 15.14 Cutaway of the refurbished building with sustainability measures

urce: Tim Soar/Cullinan Studio

sealing reduced background ventilation and a pressure test showed an airtightness value of 5 m³/h/m². All the windows are timber with aluminium facings, double-glazed with low-emissivity (low-e) coatings and argon fill.

Space heating

Having insulated to this level, and provided as much natural light and ventilation as permissible within the constraints of the existing building, low-energy technologies were then employed to provide the space heating. The building uses an air-source heat pump, located within the overrun of the old lift shaft, to provide domestic hot water and underfloor heating. The original proposal was to use a water-sourced heat pump using water from the adjacent canal, but negotiations with the British Waterways proved it would be difficult to gain consent, despite a previous connection to the canal that had provided firefighting water.

Natural ventilation

The building is naturally ventilated due to adequate cross-ventilation using openable windows, with mechanical ventilation with heat recovery fitted in the shower and toilets.

The mechanical systems are controlled by a BMS with Web pages available to monitor energy performance and space temperatures.

Lighting

The building was modelled using Relux and the predicted energy use showed the majority being consumed by lighting and IT. Task lighting is supplied at desk level with low background lighting. Computers are programmed to cut off after one hour of non-use. There are passive infrared sensors in the circulation spaces and toilets to operate the lights.

Renewable energy

The original planning permission required 10 per cent of the building's energy to be provided from on-site renewable sources, in line with Greater London Authority's requirements at the time. Five building-mounted wind turbines were originally to be used to satisfy this requirement but subsequent to the granting of permission, testing of the local wind regime and further analysis indicated that this would not be a sensible or economic solution. An area of 22 m² of photovoltaic (PV) panels on the south slopes of the roof generating electricity were chosen as an alternative and agreed with the planners.

Other sustainable refurbishment

One objective of the refurbishment was to reduce the embodied energy in the project and this was achieved by reusing most of the original structure, resulting in only one-third of the materials of the completed project being new. Cellulose fibre insulation has been used where possible and the use of PVC has been minimised.

Indoor cycle racks for twenty cycles have been provided, together with a shower.

Energy reduction and environmental achievements

The modelled total energy use was 70 kWh/m²/year emitting 32.1 kg CO_2/m²/year. Since the building was occupied, the estimated energy consumption for a whole year, based on initial meter readings, is 115 kWh/m²/year, equivalent to 59.6 kg CO_2/m²/year. This high figure is thought to be due to setting up the building and to a number of issues with the control of the heat pump. The commissioning of the BMS has been an ongoing issue. Further effort is being put into reducing the consumption.

The building achieved a BREEAM 'Excellent' rating.

Finance

The cost accounting for the project was complicated as the whole conversion and sale of the adjoining site for the construction of the new flats was an agreed package with the contractor. The cost of the conversion works is estimated as £1,720/m² for the 785 m² offices. It is not possible to calculate the additional cost of the sustainability measures over and above those

Figure 15.15 The new northern facade

Source: Tim Soar

required by planning and Building Regulations and as the project was conceived as an integrated environmental refurbishment.

Getting the building into operation

Cullinan Studio intends to carry out a post-occupancy evaluation after the first year of occupancy. The project has served as research for the practice to test ideas on designing low-energy buildings and it is the intention to continue to monitor energy and water use, and fabric performance.

Since 2008 the practice has worked to reduce its overall carbon footprint, achieving a 39 per cent reduction in three years and hopes to make a further large step with their new offices. The practice also believes that the open-plan layout, with view between floors and a shared kitchen, aids the flow of conversations and ideas between colleagues working in the Foundry.

There is ongoing monitoring of the energy use via the BMS and sub-meters and work is being carried out in the defects rectification period to reduce the higher than expected energy use.

Being the architect, client and user has given the practice a deeper insight into the process of commissioning and maintenance of a low-energy building, by providing an understanding of building–user interaction problems. It concludes that the fragmented structure of the construction industry must be addressed in relation to separate mechanical and electrical (M&E) sub-contractors, BMS programmers and installers, which complicates commissioning and compromises performance.

Figure 15.16 The single-volume, two-storey design studio
Source: Tim Soar

CASE STUDY 4: THE ANGEL BUILDING, LONDON, UK

Large 1980s office building completely renovated, retaining structural frame

- new facade clad in a high-performance, double-glazed system with solar shading;
- restaurants and retail units on the ground floor;
- five-storey atrium at the centre of the building;
- 40 per cent increase in floor space;
- displacement ventilation, low-energy lighting, low-energy pumps and lifts;
- biomass boilers and photovoltaic roof arrays;
- social survey shows wide area benefits;
- BREEAM 'Excellent'.

Figure 15.17 The Angel Building, now a thriving hub
Source: Tim Soar/Derwent London

The Angel Building is the reinvention of an unloved early 1980s commercial building located on one of London's historic focal points where City Road and St John Street meet Pentonville Road and Islington High Street. Known as 'The Angel Centre', the building was conceived in the late 1970s and completed in 1981 when it was occupied by British Telecom for several decades until they surrendered their lease in 2006.

The departure of British Telecom highlighted many problems with the building, such as outdated servicing, inefficient layout and deteriorating fabric, which meant it would be impossible to attract a new tenant without significant investment. To compound this, the building had not aged well, and was unpopular with the local population who felt it detracted from the area. After some initial studies, a major redevelopment was initiated by owners Derwent London.

Derwent London took the tired 15,000 m^2 building, set back from the road and hidden behind trees and shrubs and transformed it into a thriving hub. They added substantially more floor space and reconnected the frontage with the streetscape. A striking new facade was clad in a high-performance, double-glazed system with solar shading, and restaurants and retail units on the ground floor. The completed building attracted a number of awards, including being shortlisted for the prestigious RIBA Stirling prize.

Figure 15.18 Aerial view of the site
Source: Kevin Allen/ Derwent London

Figure 15.19 The original building
Source: Tim Soar/Derwent London

The refurbishment drive

Client: Derwent London
Architect: Allford Hall Monaghan Morris
Structural Engineer: Adams Kara Taylor
Project Manager: Buro Four
Cost Consultant: Davis Langdon
Main Contractor: BAM Construction
Local Authority: Islington Building Control
Services Engineer: Norman Disney & Young

Early analysis of the existing building suggested it offered a number of opportunities, the in-situ structural frame was sound and had good floor-to-ceiling heights implying reuse was viable. The open courtyard in the centre of the building and large spaces to the perimeter suggested there were opportunities to increase floor area to finance the recladding and general reconfiguration without wholesale demolition. In total, the new building has added about 9,200 m². The result is an essentially new building, with 40 per cent more useful floor space, which retains and extends the structural frame of the old one. The refurbished building was opened in October 2010.

A forgotten service courtyard has been transformed into a grand top-lit atrium, complete with break-out areas and a cafe. Extending to almost 885 m², the atrium is now at the core of the building, acting as a central hub. The new atrium is five-storeys high, approximately 25 metres and has an ethylene tetrafluoroethylene (ETFE) roof, allowing natural light to flood the space and into the offices around.

The architect AHMM's aims for the project included:

- creation of a landmark building with a clear identity;
- creation of efficient, comfortable and desirable office spaces;
- enabling multi-let with horizontal and vertical split capabilities;
- creation of active and viable retail spaces at street level;
- production of a 'new building' using existing components including the frame;
- regeneration of the public realm to create new external city spaces;
- integration of new green spaces with the building architecture;
- use of an integrated energy strategy to create an energy efficient building;
- achievement of BREEAM 'Excellent';
- achievement of best practice sustainability.

Whilst the external cladding, services and internal finishes of the existing building had reached the end of their life, the reinforced concrete structure proved to be sufficiently robust and with suitable floor-to-floor heights (approximately 3.7 m) to make retaining and reusing it a possibility.

The Derwent London 'Sustainable Framework for Developments' was followed from the start of the project. This ensured that all aspects of sustainability would be brought together and considered throughout the project life cycle.

Energy efficiency

Energy-efficient performance was considered essential to delivering a high-quality building. The Angel Building does this in three major ways: first, by minimising embodied energy in its structure; second, by being equipped with a full range of energy-saving measures and renewable energy technology; and third, by being well-placed for public transport and being cycle-friendly.

Reusing the building frame

Analysis of the embodied energy contained within the structure confirmed that not only would reuse realise cost savings but also significant CO_2 savings. Avoiding the demolition and disposal of the structure and construction of a new replacement resulted in CO_2 savings equivalent to 7,600 tonnes, which contribute considerably to the sustainability of the project. The build time for the project was also substantially reduced by reusing the structural frame, and overall cost savings were significant. These benefits easily outweighed the added complexity of coordinating structural and building services elements which often proves challenging in projects such as these.

Figure 15.20 The building frame retained

Source: Rob Parish/Derwent London

Facade design

The new facade of the Angel Building is a high-performance double-glazed, dark grey aluminium-framed curtain walling system with metal fins and spandrel panels. Careful consideration was given to minimise the effect of solar overheating to the office areas by providing a very high level of solar control. This was achieved by combining high-performance glazing with fritting to reduce solar gain. In addition to ensuring the necessary compliance with the Building Regulations thermal performance criteria, this has also enabled a displacement cooling system to be installed that offers high occupancy comfort combined with low energy demand. The overall insulation value of the facade is significantly better than the current Building Regulations demand. The new cladding features large motorised windows measuring 3 m x 3 m instead of the more usual 1.5 m x 1.5 m. Opening windows allow tenants maximum flexibility in their choice of air handling, naturally ventilated or (with the windows closed) mechanically ventilated, cooled and warmed. In addition, the larger windows were integrated into the design to allow tenants to have views of the mature trees on all sides.

Lighting

This is known to be one of the most important contributors to lowered energy demand in offices. Lighting controls are provided to ensure efficient operation and avoid unnecessary use. The lighting design incorporated high-efficiency fittings that aim to exceed the requirements laid out in Part L2A of the Building Regulations, and circuits set out in order to allow the use of daylight at the perimeter via daylight sensor controls. All luminaires are provided with high-frequency electronic control gear.

Ventilation

The office areas are ventilated and cooled using a displacement ventilation system. The system includes roof-mounted air handling units feeding each core. This displacement system provides the opportunity to use outside air for cooling when suitable external ambient conditions permit. This is commonly referred to as 'free cooling', as the chiller plant is not required to operate during these periods. Free cooling is estimated to be available for approximately 80 per cent of the building's standard operating hours.

Low-energy lifts

'Otis Gen2' lifts are made from recycled materials, with an intelligent 'flux vector' control system which responds to load and speed. Using 50 per cent less energy than conventional lifts, these generate energy from braking which is fed back into the building's electrical system. They are also very smooth-running.

Chillers

The building cooling needs are served by two water-cooled chillers with high seasonal efficiencies.

Pumps

Variable speed controls have been used for the heating and cooling water circuits which circulate only the volume of water required to match the required load and for correct plant operation, giving an estimated saving of between 66 to 86 per cent of traditional pumping energy use.

Low carbon and renewable energy

A renewable energy contribution was achieved using a biomass installation to produce hot water for heating and the hot water demand. Two wood pellet biomass boilers provide the majority of the heating demand, reducing dependence on gas. The ash from the boilers is then collected and composted on site and used as a soil enhancer for the landscaping and allotment area.

One hundred and forty photovoltaic panels, each with a rating of 245 W, have been installed on the roof, which are calculated to generate on average over 20 MWh of electricity per year.

Public transport and cycle provision

The building is now fully occupied with around 2,500 people working in it. Over 90 per cent of the building's occupants arrive at the building by some form of public transport, with increasing numbers choosing to arrive by bicycle. To cater for these expectations, a number of measures were introduced:

- a maximum of six car parking spaces (including wheelchair accessible spaces) and charging points for electric vehicles;
- secure parking for well over 300 bikes;
- shower rooms in the basement area and wet rooms on each floor for showering.

By virtue of its location, the Angel Building also has excellent access to public transport with a bus stop directly outside the main entrance and Angel tube stop two minutes' walk away.

Figure 15.21 New external landscaping
Source: Lee Mawdsley/Derwent London

Other environmental aspects

Many other environmental solutions were included in the refurbishment.

Low-impact concrete

The new concrete, as used in the atrium, used a 36 per cent PFA (pulverised fuel ash) mix, a by-product of coal-fired power stations which reduces the amount of cement used, thereby reducing its embodied carbon footprint. It also improves its aesthetic appearance. The concrete was also batched at a plant less than a mile from the site.

Water savings

The wide roof and terrace areas of the Angel Building are ideal for catching rain and that is just what they do. The harvested rainwater is filtered and used for toilet flushing, window cleaning and bin washing. The toilets and taps are water-efficient and urinals are waterless. All in all, the building saves the equivalent of 455,000 WC flushes per year.

Timber

The building used only timber from legal and certified sources, according to the Derwent London standard policy.

Biodiversity

Existing mature trees have been retained and new semi-mature trees and a meadow habitat have been added to the landscape around the building, all designed to create an agreeable micro-climate between streets and building. Bat flight corridors are protected. A tenants' 'Allotment Society' has also been created.

Getting the building into operation

Derwent London has its own 'Sustainability Framework of Asset Management' which has been put into operation in the buildings.

Environmental achievements

Energy

The building was designed to achieve carbon emissions of 33 kg/CO_2/m^2 and with the energy-efficiency measures it achieved a calculated carbon saving of 20 per cent below the current Building Regulations. The biomass heating installation gives the building a 15 per cent renewable energy contribution to total energy demand.

Environment

A BREEAM 'Excellent' rating was achieved for the building design.

Social aspects
Following the final letting at the Angel Building, Derwent London commissioned a study to evaluate the socio-economic impact of the regeneration of the building on key local stakeholders.

Fostering economic prosperity
The study indicated that the building, which had been vacant for a number of years, now held around 1,700 (now 2,500) employees, who each spent an average of £620 per year in the vicinity. According to the report this led to a 19 per cent increase in revenue to local businesses. In addition during construction around sixty new local jobs were created or supported in the local area.

£935,000 was invested in the public realm through planning control requirements (section 106 payments). This included increasing the number of trees and opening up the area in front of the building, providing an attractive pavement area.

Well-being for occupiers
The study found that employees felt 50 per cent more engaged and positive and enjoyed work relationships 20–25 per cent more than in their previous buildings. Light public spaces, well-designed informal spaces with well-chosen artwork all contributes to this sense of well-being.

Enhancing local communities
According to the study, local residents' well-being has increased by up to 5 per cent since the building was completed, given the improved accessibility and economic mix that the building provides. Crime in the immediate area has fallen by over 35 per cent.

Awards
The Angel Building has won many awards for the quality of its architecture and design.

- 2011 RIBA Stirling prize shortlist;
- AIA Excellence in Design Award;
- BCO National Refurbished/Recycled Workplace Award 2011 and the BCI Judges' Special Award 2011;
- BCO Refurbished/Recycled Workplace Award 2011;
- RIBA London Award 2011;
- Rejuvenation category of the Concrete Society Awards 2011;
- Offices category of the 3R Awards 2011;
- New London Architecture Award 2011.

Figure 15.22 View of the atrium from above
Source: Adam Jacobs/Derwent London

CASE STUDY 5: THE NVE BUILDING, OSLO, NORWAY

General retrofitting of 1960s listed heritage offices

- first heritage renovation to achieve Class B, 120 kWh/m2 total energy demand;
- newly connected to district heating network;
- moveable external solar shading added;
- great improvement in lighting and internal comfort.

Figure 15.23 The refurbished building

Source: Hilde T. Harket/NVE

The headquarters building Middelthuns gate 29 was constructed between 1962 to 1964 for NVE, the Norwegian Water and Energy Resources Directorate. It contains a basement, lower ground floor, six office floors and a smaller seventh floor. The building is a national monument and cultural heritage building of post-war Norway. The partly architecturally listed and protected building was in need of total rehabilitation, including modernisation, energy-efficiency improvement, environment upgrading in general, whilst dealing with the protection restrictions and retaining the existing architectural quality. The aim was to show that it is possible to combine all these aspects in a comprehensive upgrade.

The protected features of the building included:

- some details of the architecture and internal components;
- external facades and internal walls;
- the canteen, main entrance, main staircase and two of the office wings.

The building owner is Entra, and the NVE is the tenant who renegotiated the leasing contract. Both parties agreed that the building was in need of renovation and a more flexible layout with more open solutions. Both parties were committed to high energy-saving targets.

Figure 15.24 The building before refurbishment
Source: Rune Stubbrud/NVE

Owner: Entra eiendom
Architect: Dark Arkitekter AS
Consultants: Erichsen & Horgen AS, Multiconsult AS and others
Partner: Directorate for Cultural Heritage
Main contractor: Skanska

Refurbishment drive

The decision to renovate the building was taken in 2008 and the work was completed in 2011. Structurally, the building was in good condition, being built to high standards using good materials, but the indoor climate was considered to be poor. The occupants were moved out for the renovation works and housed temporarily in a rented building for two years.

The building is protected by the Norwegian Directorate of Cultural Heritage who participated in the entire process and helped decide on acceptable goals and solutions. Great care was taken to preserve the historic significance of the building with all parties (conservation, economic, visual, etc.) represented at all the meetings.

The energy ambitions were upgraded from class C to class B during the project period because B turned out to be within reach due to better airtightness. The building was the first renovation in Norway to achieve the class B standard.

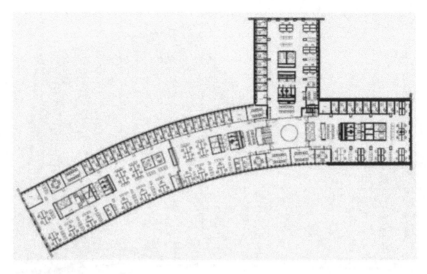

Figure 15.25 Building plan
Source: Entra eiendom

Sustainable refurbishment

Energy

The energy use of the building was modelled using 'Simien', a Norwegian simulation program. The program was used to model the numerous energy alternatives for the refurbishment.

The insulation of all the external elements of the building was improved as part of the works. The insulation of the roof on the sixth and seventh floors was increased, raising the U-value from between 0.78 and 1.12, to 0.26–0.13 W/m^{2}K. To improve the external walls, insulation and sealing measures were added under and between windows. The walls under the windows, which included protected teak panels, were updated with added insulation behind the radiators, raising the U-value to 0.11–0.31 W/m^{2}K. Between the windows, the narrow supports were insulated by removing the teak cladding and adding mineral wool behind. New krypton-filled double-glazing units were fixed into the existing teak window frames with some minimal adjustments, replacing the original double-glazing. This gave a total U-value with the frames of 1.3 W/m^{2}K. Weather-stripping was fitted around all the windows. Insulation was added under the cellar ceiling giving a U-value of 0.15 W/m^{2}K. Mineral wool insulation was used in all locations.

New external movable shading devices were installed on all the external facades to reduce solar gain. These are controlled by on-roof sensors with override available in the meeting rooms to allow for different ambiances.

Due to the narrow building plan and large windows all office areas have good daylight conditions and this was complemented by installing new energy-efficient lighting with daylight control in all offices.

Figure 15.26 Inside windows after renovation
Source: Rune Stubbrud/NVE

The space heating was originally supplied by direct electric heating using fixed panel heaters beneath the windows. During the refurbishment, the building was connected to the local district heating network and heat supplied via radiators under the windows in the offices as well as to the ventilation air. A new VAV ventilation system was installed using ducts and technical installations in the centre of the building cross-section, with no crossing of ventilation ducts possible, due to low ceiling height. The air intake for the ventilation system was on the roof and a new intake was constructed outside in the garden, with air drawn through ground ducts to provide heat-tempered air. A heat recovery system was installed on the extract air with an efficiency of 80 per cent and night cooling was introduced to take advantage of the thermal mass of the building, particularly the ceilings. The Specific Fan Power of the system was calculated as 1.57 W/m^3 at 80 per cent air flow (calculated according to NS3031). The ventilation is controlled by occupation levels via CO_2 level sensors.

To reduce cooling demand, the NVE agreed to change their indoor climate requirements and accepted more hours with temperature above 26°C degrees.

Domestic hot water is supplied from electric decentralised water heaters on each floor.

Studies for the use of renewable energy sources were carried out but none were considered appropriate. Due to the availability of the district heating system, the ground-sourced heat pump option was rejected.

Other environmental improvements

The Norwegian code for new office buildings was used for the overall design standards of the refurbishment.

The existing teak doors were reused as new doors or for other applications. All new materials were chosen for their good environmental profile, using documentation through a system called BASS. All woods came from sustainable forestry and there was no use of new tropical hardwoods; all aluminium was minimum 30 per cent recycled and steel 50 per cent recycled; similarly, all products for interior fittings and furniture were environmentally certified.

New sanitary equipment was chosen and installed with high water-efficiency standards.

The on-site refurbishment works were carried out in an environmentally friendly way with 85 per cent of the building waste separated on site, and all hazardous materials to be demolished were identified in advance via an environmental redevelopment methodology.

Energy reduction

This was the first protected building to be renovated to energy level B or better. Prior to refurbishment the measured overall energy use for the years 2006–8 (temperature compensated to normal year) was 213 kWh/m²/year, including all equipment. After renovation, the calculated overall delivered energy demand is 119 kWh/m²/year, compared with the regulations for new buildings of 150 kWh/m²/year, including the net space heating demand of 36 kWh/m²/year.

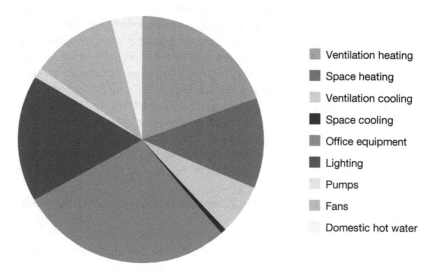

Legend:
- Ventilation heating
- Space heating
- Ventilation cooling
- Space cooling
- Office equipment
- Lighting
- Pumps
- Fans
- Domestic hot water

Figure 15.27 Calculated energy use

Source: Entra eiendom

The airtightness post-refurbishment was measured as 0.8 air changes per hour.

Financial aspects

The total cost of the renovation works was 152 million NOK (Norwegian kroner) excluding VAT: less than 20 million euros.

Financial support came from ENOVA, the Ministry of Energy and Petroleum, for conversion from direct electric heating to district heating. The remaining financing was provided by Entra on the basis of a 50/50 loan and equity capital. Entra has a requirement for all its investments to be commercially profitable.

The building in operation

The occupants are now very happy with the building, particularly as it is much lighter with much more daylight than before, due to the opening up of the internal spaces.

Commissioning the solar shading system and controls caused some initial difficulties, as did the setting up of the BMS.

The building and its renovation have been a great success story and the owners are proud that they now have so many visitors.

Figure 15.28 The building retrofit is deemed a great success

Source: Hilde T. Harket/NVE

CASE STUDY 6: CENTRO COLOMBO SHOPPING, LISBON, PORTUGAL

Refurbishment of major shopping centre to generally upgrade and improve sustainability aspects

- windows replaced to balance solar gains with daylighting;
- all lighting upgraded as necessary;
- building services upgrade, based on CHP system;
- new roof garden created;
- water conservation measures installed;
- 100 per cent compliance with the critical standards in Sonae Sierra's tool – 'Environmental Standards for Retail Developments' (ESRD).

The Colombo Shopping Centre, located ten minutes from the centre of Lisbon, opened in September 1997 and covers a land area of 85,000 m^2 with a total floor area of 120,000 m^2. It is one of the largest top ten shopping centres in Europe and contains shops, restaurants and food centres, a health centre, cinemas and 'funcentres' as well as a chapel and police and post offices. The centre has been fully occupied since it was opened, with nearly 25 million visitors in 2009 spending more than 444 million euros. Sonae Sierra are 50 per cent owners and 100 per cent managers of the centre and embarked on a renovation of the centre in 2008 with works completed in March 2009.

Refurbishment drive

The renovation works were part of regular upgrading of facilities at the centre and focused on several specific areas, including: the food-court, sanitary installations, an exterior garden created on the roof, installation of all new lifts, several improvements in the parking area and other specific improvements all around the centre.

Sonae Sierra had previously developed their 'Environmental Standards for Retail Developments' (ESRD), an internet-based specification tool designed to support development managers and design teams in providing retail developments that comply with the company's environmental principles and Environmental Management Systems (EMS) procedures. This was used on the Colombo refurbishment project. The ESRD sets out 190 standards, divided into critical and leadership standards, covering environmental issues including energy, water, waste, transport, health and well-being, site sustainability and materials. The standards encompass the best

Figure 15.29 Inside Colombo
Source: Sonae Sierra

environmental practices in the planning and design of shopping centres and were developed based on Sonae Sierra's experience, the best available techniques in the market and existing building environmental assessment schemes such as Leadership in Energy and Environmental Design (LEED) and BREEAM.

The objectives of the refurbishment were to upgrade the shopping centre and improve the environmental performance. This included assuring conformity with current environmental legislation and regulation, adopting responsible standards in cases where legislation is incomplete or non-existent and enabling operation of the centre in an environmentally responsible way.

The following areas were prioritised for improvement:

- climate change mitigation and adaptation;
- health and well-being;
- water efficiency;
- materials;
- efficient space use with biodiversity.

Energy refurbishment

The centre already had a combined heating and power (CHP) system but certain elements of the structure, the HVAC system and lighting systems needed upgrading to conform to the ESRD.

The old panels on the periphery of the skylight of the food-court were replaced by new vertical glazing with decorative features made out of a special PVC film with polymeric non-migrating plasticiser. The purpose was to avoid condensation on internal surfaces and reduce solar heat gains. The new glazing facade has a U-value of 2.7 W/m^2K (better than the environmental standard 2.9). Glazing selection was based on the best compromise between solar gain control and daylighting penetration.

To decrease thermal losses in the HVAC distribution system, the insulation of pipes and duct was replaced using different types of insulation and thicknesses according to the pipe diameter, water temperature and duct dimensions and locations. A free-cooling strategy was introduced in the ventilation systems to decrease energy use for active cooling.

Two new air-handling units (AHU) serving level two of the food-court were fitted to increase capacity as well as the refurbishment of two other existing units serving level one. The new AHUs are now operated by indoor air quality-based controls via the BMS, and include heat recovery wheels with an efficiency of 55 per cent. The fans in the AHUs were replaced with high efficiency motors, Eff 1 class.

To monitor and control energy use and to enable continuous improvement of the energy performance of the centre, the existing BMS was improved to control the free cooling, IAQ and optimum start and stop control on the heating.

Lighting refurbishment works were carried out in the food-court areas and installed power is now 23 W/m^2. Electric circuits were also redesigned to reflect the different lighting zones inside the food-courts, giving three circuits providing different lighting levels in each zone. Any non-electronic ballast was replaced. All lighting circuits have automatic time clock control with manual override and the ability to implement different schedules for each day of the week and to change hours of the circuits from one week to the next. The BMS implements the minimum lighting control. In zones that benefit from daylight, lighting control strategies use time clock control based on daylight duration or photo-sensors.

One of the shopping mall entrances giving access to the roof garden was rebuilt to provide a vestibule entrance with self-closing double doors

Figure 15.30 Lighting improvements
Source: Sonae Sierra

distanced more than 4 metres apart, allowing the first to close when the second door opens, reducing energy losses and draughts.

Other environmental refurbishment

Health and well-being

The air quality both indoors and outdoors around the centre were considered to be very important and several studies were carried out and reports compiled, notably:

- an indoor air quality report in 2007, where levels of contaminants including particulates (PM10) , CO, CO_2 ,O_3, CO, NO_2 and SO_2, were assessed using official data from existing monitoring stations and on-site measurements;
- annual reports (2006, 2007 and 2008) of gaseous emissions related to the on-site internal combustion engines from the cogeneration plant, i.e. CO, NO_2, NO and SO_2;
- a study in March 2008 of the dispersion of pollutants in the atmosphere with a view to the future replacement of the three fuel-oil engines operating the cogeneration plant, by two new gas engines.

One of the direct results of the studies was to install new comprehensive air filters into the air intakes and ventilation systems.

The design of the AHU and related ducts was scrutinised to minimise contamination of ventilation air with micro-organisms and other contaminants. The following principles were used:

- the inside surfaces of AHUs and ducts to be smooth and non-adsorbent to enable disinfection and cleaning, the use of flexible ducts limited to the minimum, i.e. only in terminal connections;
- all filters to be class F5;
- all connections to be on the same side of the AHU to facilitate cleaning;
- drain pans to be provided beneath all condensate-producing heat exchangers and all dehumidifying cooling coils with adequate drainage arrangements.

Three smoking lounges were refurbished as part of this project, two located in the food-court and one in level zero, to give the airflow rates greater than 90 m^3/h/person. (The regional percentage of smokers in Lisbon is 17 per cent but smoking is restricted to the centre's smoking lounges.)

The ventilation rate in the food-court, with a design occupancy of 1 m^2/person, was set at 52 m^3/h/person. In the WCs the exhaust airflow rate was designed as 120 m^3/h/maximum number of occupants, with internal negative pressure, eliminating air circulation to other spaces.

Figure 15.31 Indoor air quality has been greatly improved

Source: Sonae Sierra

Water

To evaluate the efficiency of the shopping centre's internal water supply network, water meters are installed in different parts of the network, connected to the BMS. As part of the refurbishment, new water meters were installed for the irrigation system, the WCs and wash hand basins and toilets/urinals, and the ornamental fountains. All customer toilets were equipped with electromagnetically operated taps using infrared sensor, with a nominal flow rate of 6 litres/min and WCs were equipped with the similar electromagnetically operated flushes. The flow rate of the WCs flushes was reduced to 4 litres/flush. Some water-free urinals have also been fitted.

The fountains and other water features are equipped with recirculation water systems that promote the reduction of water consumption and a water treatment system comprising filtration, chlorination and redox potential measurement device.

Materials

The project team identified, quantified and registered the main materials used in the refurbishment that could be a source of air contamination, with the following results:

- Composite wood – 60 per cent were low emission with a AITIM (Asociación de Investigación de las Industrias de la Madera) classification as low formaldehyde content (E1) according to EN 120.
- Paints – 57 per cent of paints and coating products were low-VOC (volatile organic compound) emission type.
- Carpets – carpets were tested by a qualified laboratory (LQAI) which concluded that the carpet met all the parameters established by the 'European Collaborative Action, Indoor Air Quality' and its impact on man (ECA-IAQ).
- Adhesive and sealants – the project achieved 42 per cent of low-emission products.
- Insulation – no manmade fibre products were installed inside the centre.

Wood products for flooring, wood panel finishing, windows, glazing framing and doors, furniture and wood-based decorative elements were selected based on sourcing from sustainable managed forests where possible. In the Colombo refurbishment, certificates based on the Programme for the Endorsement of Forest Certification (PEFC) were presented for more than 50 per cent of timber products installed.

Alternatives to PVC materials were actively sought and analysis showed that the refurbishment achieves a 'PVC alternative solutions rate' of 91 per cent.

About 200 kg of refrigerant fluids were installed during the refurbishment. To avoid high environmental impacts the refrigerants used have

Figure 15.32 Flooring used alternatives to PVC

Source: Sonae Sierra

Global Warming Potential (GWP) less than 2,000 and Ozone Depletion Potential (ODP) of zero.

Sonae plan to maximise the use of local and regional materials in all work and in this refurbishment 95 per cent of heavy materials were regionally sourced. Sonae also promote the inclusion of off-site prefabricated elements in the construction to maximise the manufacturing efficiency of construction elements, minimising waste arising on site as well as the volumes of raw materials used during the construction phase. Based on the total cost of the building works, the component prefabrication rate for this refurbishment was 56 per cent.

Land use and biodiversity

As part of the refurbishment, a new roof garden of more than 1,500 m² was created. A landscape architect was employed to develop a landscape project including an assessment of soil quality and climatic characteristics, as well as an evaluation of the total water needs for irrigation and the specification of well-adapted plants to reduce water needs. A total of 5,000 plants of thirty-five different species have been planted in the roof garden. Two different irrigation systems were installed for the landscaping areas, drip irrigation and automatic sprinkler systems, incorporating electro-valves to control zoning, and irrigation programs and rainwater sensors to control the irrigation levels. Lighting in the garden uses light-emitting diode (LED) systems framed in vegetation.

Figure 15.33 The new roof garden
Source: Sonae Sierra

Other elements of the ESRD

Several other components of the Sonae Sierra's 'Environmental Standards for Retail Developments' were not applicable to the refurbishment, particularly waste, transport, management and *legionella* related to firefighting systems.

Energy reduction and environmental achievements

The compliance level with the Sonae Sierra's Environmental Standards for the Retail Development achieved in the refurbishment was 100 per cent for the critical ESRDs and 90 per cent in the applicable points related to the leadership standards.

In the project a recycled level of 73 per cent of all debris produced in the construction works was realised, against Sonae's target for refurbishments of 65 per cent.

The annual energy consumption of the mall and toilet areas, excluding the tenants' own consumption, was estimated to be reduced by the works from 718 kWh/m^2 in 2007 to 676 kWh/m^2 in 2009.

Financial aspects

The total investment cost of the refurbishment was 27 million euros but no breakdown of the sustainability part is available.

The building in operation

All works in this refurbishment were performed at night and without closing a single store. Both the physical upgrading of the mall and the compliance level achieved with the Sonae Sierra's 'Environmental Standards for the Retail Development' were considered to be exceptional.

The level of satisfaction regarding the accessibility between floors was greatly increased due to the increased capacity of the lifts and the bigger doors.

The HVAC system performance in the food-court area has improved considerably, which was one of the main complaints by the clients before the refurbishment. Also the natural lighting in the food-court improved significantly, as well as its overall image, together with an improvement in seating capacity of 50 per cent. The new HVAC system for the toilet areas has been a huge improvement, both in image and air quality.

The provision of the exterior garden, whose lack was one of the limitations of Colombo before refurbishment, has been much appreciated for its quality and good environment. The level of service in the parking area has improved due to the ceiling being painted giving much more clarity without any additional lighting. A new parking-help system was installed and the circulation and the interior signage of the parking improved significantly, reducing another of the worst complaints from clients before refurbishment.

There were no major difficulties with the refurbishment except that it was difficult to obtain information from the 'as-built' drawings and much of the information was only available in paper format.

CASE STUDY NO 7: AS SOLAR CORPORATE HEADQUARTERS, HANOVER, GERMANY

Derelict factory transformed to multifunctional energy-plus headquarters building

- offices refurbished to Passive House standards;
- optimised use of daylight and solar gain;
- 268 kW peak PV installed on the roof;
- solar water heating with large storage tank;
- biomass boilers;
- solar absorption chillers.

In June 2011, the AS Solar company moved into its new corporate headquarters on Nenndorfer Chaussee in Hanover. The multifunctional energy-plus building was created from a derelict factory building. The 200 employees in the AS Solar Group in Hanover carry out research, development and marketing of products globally in the PV, solar thermal and biomass heating sectors.

Owner: AS Solar GmbH, Nenndorfer Chaussee 9, 30453 Hannover, Germany
Tel.: +49 511 475578-0
Fax: +49 511 475578-11
Email: info@as-solar.com
Internet: www.as-solar.com (accessed 2 November 2013)

The building complex originally formed part of the Telefunken company's Plant II factory, which was built in 1959. Television sets were produced here until the beginning of the 1980s. The industrial building later housed a printing plant and was then used by logistics companies as a warehouse and car park until the year 2000, since when it has remained empty. The building had suffered from severe vandalism and was considered for demolition.

However, AS Solar had other plans for the building complex; they wanted to house offices, production and storage areas in the building for their own use, following a comprehensive renovation earmarked to begin in 2010. The building concept includes several innovative measures and technologies with the energy supply supplemented with renewable energies. Overall the holistic approach has managed to greatly reduce the primary energy requirement and the whole building achieves an energy usage standard that considerably exceeds the current EnEV requirements for both existing and new buildings.

Figure 15.34 The near-derelict building before renovation
Source: © AS Solar GmbH

Refurbishment drive

Because the whole area was a former industrial centre and bombed during the war, before renovation there needed to be a search of the whole site for unexploded bombs; none were found. The refurbished building is free-standing and consists of two sections constructed using a reinforced concrete frame structure, each with a gently sloped gable-roof structure. In addition to production and storage areas, 7,000 m² of office are included. All office and similar areas have been refurbished to Passive House standards and the refurbishment design optimised with a view to using renewable energies where possible. The use of daylight and passive solar gain were increased by providing not only more window area but also a large atrium located centrally in the floor plan. The daylight usage is further optimised and the requirement for artificial light is reduced by arranging light-deflecting devices in the facades. In addition to an industrial-sized kitchen, cafeteria, two seminar rooms and a technology showroom on the ground floor, office spaces, toilets, showers and recreational spaces are located on the intermediate level of the building. The production area, which has been constructed according to EnEV 2009, is directly next to the office area on the intermediate level.

Energy refurbishment

Energy modelling was carried out using TRNSYS.

Insulation

The building envelope was refurbished to Passive House standards and the ventilation system designed with highly efficient heat recovery. This has

Figure 15.35 The refurbished headquarters

Source: © AS Solar GmbH, Tom Baerwald

reduced the heating energy requirement by more than 90 per cent from around 270 to 25 kWh/m^2/year. The insulation of the office and production area walls consisted of wood fibre board between 200 and 300 mm thick, giving a U-value of 0.14 W/m^{2}K. An oriented strand board (OSB)-layer placed behind the insulation provides good airtightness. A Perlite ballast/filling was used to close gaps between the old and very irregular reinforced concrete elements.

A new profiled sheeting roof was installed directly onto the old roof. The old roof was covered by a moisture barrier layer and 300 mm of Isofloc insulation blown in, giving a U-value of 0.11 W/m^{2}K. The floor between basement and first floor was insulated by fixing fibreglass mats of 150 mm thickness, from below, giving a U-value of 0.33 W/m^{2}K. The same solution was used between packaging/storing area and office area. The windows of the offices were triple-glazed giving a U-value of 0.8 W/m^{2}K.

The storage area has been refurbished without any additional thermal protection measures since it will remain unheated.

Renewable energy

The building is heated and cooled completely using renewable energies. Evacuated tube solar collectors (150 m^2) were installed on the west, south and east elevations to supplement the heating supply provided by wood pellet boilers. An existing 30 m^3 sprinkler pressurised tank was reconstructed as a solar heat storage vessel. The old sprinkler tank was covered all over with 200 mm of Rockwool and inside the tank a stratification system was installed for the hot water coming from the solar thermal secondary circuit. The total cost at 19,000 euros was only half of the estimated cost for a new storage buffer tank of the same size.

The office areas are cooled in summer using two solar thermal powered absorption chillers with a cooling output of 19 kW each. The peak cooling demand is covered as necessary by a vapour compression chiller with a cooling output of 160 kW.

A photovoltaic plant with 268 kW peak capacity has been installed on the roof. When there is sun it provides electricity for operating the building, with surplus solar power being fed into the public grid. The building ventilation fresh air intake has a bypass through the PV inverter room so that the stream of air both raises the inverter's efficiency by cooling the inverters, and the waste heat (from 8 to 30 kW) is transferred to the building, contributing to the heating. In summer the inverters are also cooled with fresh air to achieve the higher yields. In the second construction stage it is planned to install a further PV system with a 125 kW peak output on a neighbouring warehouse, and the carports at the company parking lot will also be covered by PV modules to support electric vehicles.

Heating

There is no conventional back-up heating solution as this was considered unnecessary and the mains gas connection to the site was cut. Heating is

Figure 15.36 The biomass boilers

Source: © AS Solar GmbH

supplied by the solar thermal system and four wood burners in the showroom. There are two different SHT combined burners (log and pellets) of 40 and 25 kW, a FireFox pellet burner of 35 kW and a Swedish EcoTec BioLine pellet burner of 20 kW. So far this has been sufficient for heating and the planned installation of an additional 150 kW pellet burner has not been necessary. Domestic hot water is produced by a large heat exchanger giving water at 60°C on demand.

Energy management system
RSS hardware from the FUTUS Company from Austria was installed, adapted to manage the complex systems. A large number of sensors, heat meters and a weather station for the monitoring research programme were installed. The monitoring is carried out in cooperation with the Ostfalia College in Wolfenbüttel. The system is complicated, but uses a distributed network architecture, where the servers running the management software are programmed manually by internal IT personnel, who little by little have adapted the system to the actual site conditions.

Other environmental refurbishment
In general the renovation sought to bring back into service as much of the existing infrastructure as possible. This included the old freight elevator from the basement to the packing area, the old lifting ramps to load the lorries, the old hoisting platform and a pressurised air tank of 2 m^3 now used as a cold water storage tank. In the building itself the guidelines were to use only

recyclable, environmentally friendly materials with very low or no emission of any chemicals, thus no PVC was allowed in the electric cables; the floor was made from oak from Forest Stewardship Council (FSC)-certified forests and their surface was only oiled, not varnished. The printers use a type of wax instead of toner, to prevent the dust and ozone emission. The Isofloc insulation used is made from impregnated old paper, the Perlite insulation is natural volcanic minerals.

From the rubble of demolished brick structures, a new wall was built on the southern side of the building as noise protection from the nearby major road and as a replacement for the trees that had to be felled. The trees were cut into logs and burnt in the wood burners. Being a wholesaler, the company receives a huge number of 'one-way' pallets, which are also cut up and used for heating.

A wet cooling tower was installed in the underground exhaust air channel, so when it is operational, it makes use of the dry and relatively cool air leaving the building, increasing its cooling efficiency and lowering the use of water. The site water-softening system uses an electrolytic system, with cleaning via a switch in the electric potential which dissolves the deposits quickly so that they can be flushed away in concentrated form, without chemicals.

Energy balance

Overall the building achieves an energy-plus standard for the annual energy balance due to the solar and wood heating and the large PV array. The PV array has generated an average of more than 250 MWh of electricity per year over the past four years, with the whole building using less than this per year.

Financial aspects

The costs of refurbishment were 239 €/m^2 for the construction and 171 €/m^2 for the technical services installations, based on the gross floor area.

Getting the building into operation

The building was converted in sequence starting with the roof to prevent further harm done by rainwater penetration, next was the PV array and then the warehouse followed, and so on. The refurbishment process was not difficult but several parts of the services infrastructure demanded considerable effort to commission them, including the cooling system, air-conditioning system, solar thermal installation and the pellet burners. The PV system and inverters, and the domestic hot water (DHW) system required much less effort.

AS Solar are also planning to construct a solar vehicle-charging station at the company's headquarters. The refurbishment building is being monitored to evaluate and optimise the refurbishment concept and create

Figure 15.37 The new office
Source: © AS Solar GmbH

guidelines for refurbishing commercial buildings in accordance with the energy-plus standard. It will analyse the energy savings potential that can be achieved using a coordinated selection and integration of passive and active measures, and by using optimised zoning and usage.

The focus of the refurbishment was to demonstrate the choices available today to decrease environmental impact from housing and working and to build a 'lighthouse project' to showcase future renewable energy systems running for customers. A great success has been the ongoing interest shown by visitors for conducted tours through the building.

Information provided by AS Solar and by:
EnOB: Research for energy-optimised construction

'Buildings of the Future' is the guiding concept behind EnOB – research for energy-optimised construction (the name EnOB is an abbreviation of the equivalent German term Energieoptimiertes Bauen). The research projects sponsored by the German Federal Ministry of Economics and Technology involve buildings that have minimal primary energy requirements and high occupant comfort, with moderate investment costs and significantly reduced operating costs.

For more information see: www.enob.info/en/ (accessed 3 November 2013).

CASE STUDY 8: KREDITANSTALT FÜR WIEDERAUFBAU (KFW), FRANKFURT, GERMANY

Total renovation of large 1960s office block complex

- new facade includes novel moveable shading system;
- mechanical extract ventilation, with above window air inlets;
- CHP provides heating and absorption cooling;
- lighting control optimises use of daylighting.

The headquarters of the Kreditanstalt für Wiederaufbau (KfW) was built in the late 1960s as a group of office towers in Frankfurt's west end. By 2000 the buildings were starting to look somewhat down at heel and the owner decided that they needed to undergo radical modernisation, with only the load-bearing framework from the original building remaining intact.

The four office towers were constructed between 1964 and 1968. They are staggered relative to one another by the depth of a cellular office space and rise to different heights, with a maximum height of 62 m, not unusually high buildings by Frankfurt standards.

The renovation enabled daylight, fresh air, and heating and cooling to be considerably improved and to be operated in a much more energy-efficient way. The facade, interior fittings and technical services for the 22,000 m^2 building complex were renewed between 2002 and 2006. The fire protection and workplace conditions are now also state of the art. Furthermore, with this ambitious refurbishment the KfW Bankengruppe has demonstrated its commitment to environmental and climate protection.

re 15.38 Facade before rbishment

rce: EnOB

Refurbishment planning

In terms of the fire protection regulations alone there was need for improvement, and the appearance of the towers had also long ceased to match the image which the owner and occupants wished to convey.

The KfW Bankengruppe places considerable importance on user comfort with low energy consumption for all its buildings. Hence, the new facade was intended to control the light, air and heat supply, and, together with new building services equipment, to provide primary energy savings of around 50 per cent. The refurbishment was intended to be state-of-the-art modernisation but with the office towers demonstrating much more than just a new appearance.

Figure 15.39 Blocks with the new facade

Source: RKW Carsten Costard

The fire protection requirements were of primary importance, with new escape routes designed, and separate fire and smoke compartments created in the buildings. In addition, early fire detection measures were provided and new access arrangements for the fire brigade created.

There were some limited possibilities for extending the building complex in the location. On the north side there are a number of mature trees; on the south side there is an arcade building from the 1980s which is used by the bank. This left only the west side, where the German Library (Deutsche Bibliothek) had stood until recently. Here it was possible to build an event and exhibition hall in front of the two southernmost towers. The third tower in the row was extended with a two-storey conference room. The northern tower was heightened by three full storeys and now houses the executive boardrooms. The ground floor was also extended and now has a new entrance area that looks onto the courtyard. The newly designed office areas consist of both small individual offices and team offices for groups of six to ten staff members.

The building was reoccupied in August 2006 and monitoring data have been recorded and evaluated since February 2007.

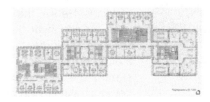

Figure 15.40 Arrangement of the original blocks

Source: © RKW Architektur + Städtebau

Energy refurbishment

The energy-based refurbishment of the office towers centred on five aspects.

The new facade

The facade includes a new kind of sun protection and aerodynamically optimised supply air features. The sun protection elements can be moved vertically up and down and can also be tilted, controlled centrally and individually for each room. The elements, which consist of metal mesh sandwiched between glass panels, allow daylight to penetrate while at the same time providing some view to outside.

Insulation

The thermal insulation of the blocks was considerably improved, moving the facade U-value from 1.1 to 0.28 W/m^2K, the windows from 2.9 to 1.4 W/m^2K and roof from 0.79 to 0.21 W/m^2K.

gure 15.41 Novel shading system
urce: © RKW Carsten Costard

Ventilation

The ventilation strategy uses customised control concept and passive cooling using night ventilation to reduce the cooling energy demand. Because of the low floor-to-ceiling heights and the numerous downstand beams, a full air supply and exhaust system would have been problematic. Thus a central air exhaust only system was fitted in the office areas with the necessary fresh air coming from openings above the windows. The exhaust air is drawn through gaps into the space above the suspended ceiling and from there into a collection duct, leading up the stairwell to where the air is vented via a central exhaust fan. The air supply is controlled according to the outdoor temperature and occupation levels, to reduce energy loss in cold weather and to avoid unnecessary heat gain in hot weather.

ure 15.42 Pressure testing the building
rce: © ip5

Cooling

As it was not possible to avoid suspended ceilings for a number of reasons, control of room temperatures is provided by cooling panels in the suspended ceilings. Since the mid-1990s there has been a CHP plant for heat supply to the KfW buildings and waste heat from this plant is now used by an absorption cooling system. In addition, for very hot days there are compression chillers for boosting the cooling supply.

Daylighting

The lighting system uses presence and daylight detection control to optimise the use of daylight in the offices. Depending on the amount of daylight available, measured by a sensor above the table surfaces, artificial lighting is provided to give the light level previously set by the user. The target light level does not remain constant but increases slightly as the external light level increases, on the basis that a room with a constant level of lighting is considered to be darker when the external daylight level increases. If the occupant forgets to switch off the lights in an office when they leave, the building control technology will do so automatically fifteen minutes after the room is vacated.

Other environmental refurbishment

Water saving technologies include grey water and rainwater collection and treatment for sanitary use.

Energy reduction and environmental achievements

Energy use calculations show that the energy use for heating has been reduced from 113 to 74 kWh/m^2 and the total primary energy use from 215 to 94 kWh/m^2.

Information provided by:
EnOB: Research for energy-optimised construction

'Buildings of the Future' is the guiding concept behind EnOB – research for energy-optimised construction (the name EnOB is an abbreviation of the equivalent German term Energieoptimiertes Bauen). The research projects sponsored by the German Federal Ministry of Economics and Technology involve buildings that have minimal primary energy requirements and high occupant comfort, with moderate investment costs and significantly reduced operating costs.

For more information see:
www.enob.info/en/
(accessed 3 November 2013).

Figure 15.43 Underfloor heating installation
Source: © ip5

CASE STUDY 9: EBÖK VERMÖGENSVERWALTUNG OFFICE, TÜBINGEN, GERMANY

Old barracks building converted to Passive House standard office

- high insulation for walls, roof and windows;
- floor slab peripheral insulation creates heat store under slab;
- ground-sourced heat exchanger provides cooling and frost protection;
- night cooling using mechanical ventilation produces comfortable conditions.

This former Thiepval Barracks building was constructed by the French garrison in 1954 for teaching purposes. It had not been used for some time and was in a very poor state of repair. Nevertheless, the local conservation authority had listed the ensemble of buildings and stipulated that the outer appearance of the buildings had to be preserved.

The location of the building, on the site of the barracks very close to Tübingen's main railway station, is considered very attractive. There are good connections to local and regional public transport networks as well as to the cycle path network. The former parade ground is now a public park and the historic centre of Tübingen can be reached on foot.

The building was in a poor structural condition when it was bought by ebök GmbH in 2002 but with the refurbishment as their new offices in Tübingen, they were able to reduce the heating and electricity requirements significantly and increase comfort in summer. It was, in fact, the first building in the world to be awarded a Passive House certificate for refurbishment.

General refurbishment

A new roof truss comprising double I-joists (TJI) was built on top of the old ground floor and with two new dormer windows so that more workspaces could be accommodated in the roof.

With the aim of achieving high-quality and comfortable workspaces with minimum energy requirements, an energy concept was chosen that mixed proven state-of-the-art technologies with prototypes and experimental measures. The objective was also to achieve Passive House standards.

The thermal insulation for the walls and the roof was carried out without difficulty and without creating any thermal bridges, but the insulation of the ground floor presented problems typical of old buildings. To increase the insulation thickness of the foundation slab would have meant raising all the

Figure 15.44 The refurbished offices
Source: © ebök, Tübingen

Figure 15.45 The original barracks building
Source: © ebök, Tübingen

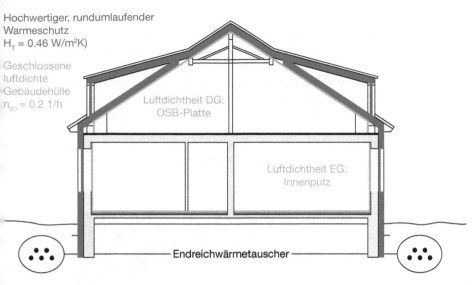

Hochwertiger, rundumlaufender
Warmeschutz
$H_T = 0.46$ W/m²K)

Geschlossene
luftdichte
Gebäudehülle
$n_{50} = 0.2$ 1/h

Luftdichtheit DG:
OSB-Platte

Luftdichtheit EG:
Innenputz

Endreichwärmetauscher

ure 15.46 Cross-section showing the insulation

rce: © ebök, Tübingen

door lintels, which was not possible, and so the solution chosen was to further increase the insulation to the external walls. This was shown by the dynamic, two-dimensional heat flow calculations to achieve a standard comparable to the Passive House requirements. Flanking insulation of the slab was also shown to have the effect of increasing the temperature in the trapped ground volume beneath the building over time, creating a 'heat sink' beneath the building reducing heat losses and leading to higher temperatures in the floor.

Energy refurbishment

Achieving Passive House standard

The total energy demand for the offices was calculated according to the Passive House Planning Package (PHPP). The Passive House refurbishment requirements are 20 kWh/m²/year for heating (space and hot water) and 7 kWh/m² for lighting and building services equipment. Converted into primary energy, this corresponds to a total energy demand of 43 kWh/m²/year. This means that the building requires only 15 per cent of the average primary energy use in existing office buildings in Germany, or 60 per cent lower than the energy requirements for the 'Energy-optimised construction' (EnOB) standard.

Ventilation

Work and meeting rooms have a ventilation system that provides fresh air, with exhaust air extracted from the ancillary spaces such as the WC, kitchenette and server room. The corridors are transfer flow zones, thus only the work and meeting rooms receive the supply air. A programmer controls the ventilation in the office spaces, but depending on the number of occupants, the volume flow can be increased using an auxiliary fan. The mechanical night ventilation used in summer is provided independent of the exhaust and outside temperatures.

The ventilation system has high-efficiency heat recovery and is a prototype that greatly exceeds the efficiency criteria for Passive House ventilation systems. In addition, preheating of the supply air and frost protection for the heat exchanger are provided by a brine flat-plate collector laid in the ground around the building.

Figure 15.47 Laying the pipes for the brine ground heat exchanger

Source: © ebök, Tübingen

Heating

The high-quality thermal insulation to Passive House standards reduced the heating requirements considerably below the statutory minimum requirements stipulated by the German Energy Saving Ordinance (EnEV).

The building is conventionally heated with panel radiators running at between 65 and 45°C using a modulating gas condensing boiler when necessary. A continuous flow gas heater provides hot water across the short distances in the kitchenette, shower and wash room.

Shading and reducing heat gains

In summer, internal solar shading is used to reduce solar gain and is supported by passive cooling. The internal shading is a compromise solution between the need to provide solar protection and the listed building requirements. Internal loads are minimised by using efficient electrical equipment and lighting. The heat from the large source, the server, is extracted directly from the equipment cabinet. Overall the measures were designed to ensure a pleasant indoor environment.

Cooling

For night cooling, the mechanical ventilation system delivers a large volume of cool air through the building so that the ceilings, which provide most of the thermal mass, release their stored heat. For structural reasons, it was not possible to construct the top floor using a thermally massive structure so the ceilings here were clad with gypsum board impregnated with PCM using micro-capsulated paraffin, produced by the manufacturer as a prototype. The brine ground-sourced heat exchanger system is used to cool the outside air for the night ventilation.

Daylighting

A simulation program was used to design a good working environment using daylight as available to give optimum lighting of the workspaces. A highly efficient, dimmable lighting system was created, in which daylight and motion sensors control the artificial lighting levels.

Renewable energy measures

Use of the roof to install active solar panels was not possible due to the buildings being listed as a historic ensemble.

Energy reduction and environmental achievements

The mean monitored energy consumption for heating and hot water during two years was 24.5 kWh/m^2/year, including generation and distribution losses. The calculated heating requirement of 22.4 kWh/m^2/year including losses forecast during the planning, is therefore almost achieved.

The electricity consumption is of the same order as the heat energy consumption. However, in terms of the primary energy, the electricity consumption dominates the overall energy consumption as a consequence of the conversion factor. This shows how important it is to install energy-efficient office equipment and building services systems in energy-saving buildings.

The technical office equipment dominates the electricity use, which at an average of 20.8 kWh/m^2/year is around 70 per cent of the overall electricity consumption.

In 2004 the building became the first refurbished building to be awarded a Passive House certificate from the Passive House Institute in Darmstadt.

Figure 15.48 Central ventilation unit with heat recovery in the roof space

Source: © ebök, Tübingen

Financial aspects

The construction and technical services costs amounted to 810,000 euros (net), or 836 euros/m². These include additional costs for purely energy-based additional measures without summer cooling and without lighting.

Without taking into account any subsidy, after thirty years this provides an excess net present value of around 20,000 euros, while the payback time amounts to twenty-six years.

In this case, a grant was awarded by the Baden-Württemberg Ministry for the Environment and Transport as part of its Climate Protection Plus Programme, which amounted to 47,200 euros. For the assessed period of thirty years, this leads to an excess net present value of around 112,000 euros, while the payback time reduces to six years.

Building in operation

Combining the ventilation and night cooling in the ceilings has had a very positive effect on the comfort in the offices, removing the internal heat gains and part of the solar gain. The mechanical night ventilation generally cools the thermal mass by a sufficient amount but there is still some room for improvement. During hot periods, there are increased room temperatures as a result of the limited output of the ground collector together with the relatively low solar shading effect provided by the internal blinds. In some rooms, the frequency of overheating amounts to around 10 per cent of the working time.

Information provided by:
EnOB: Research for energy-optimised construction

'Buildings of the Future' is the guiding concept behind EnOB – research for energy-optimised construction (the name EnOB is an abbreviation of the equivalent German term Energieoptimiertes Bauen). The research projects sponsored by the German Federal Ministry of Economics and Technology involve buildings that have minimal primary energy requirements and high occupant comfort, with moderate investment costs and significantly reduced operating costs.

For more information see: www.enob.info/en/ (accessed 3 November 2013).

CASE STUDY 10: SCOTSTOUN HOUSE, EDINBURGH, UK

Listed office building in Scotland retrofitted to high energy standards

- extensive solar tube daylighting;
- biomass boiler;
- natural ventilation with wind catchers;
- phase-change material in ceiling insulation;
- comprehensive control systems installed with energy monitoring;
- BREEAM 'Excellent'.

Scotstoun House was purpose-built for the engineers Arup in 1966 and has been owned and used by them since then. It is a low single-level structure within the walled garden of a previously demolished country house at South Queensferry, 8 km west of Edinburgh. South Queensferry is strikingly located on the south bank of the Firth of Forth, between the southern ends of the cantilever Forth rail bridge and the suspension Forth road bridge.

Refurbishment drive

By the early 2000s, the original building had started to show the limitations of its 1960s technology, and was also becoming a constraint on the way Arup wanted to work. A development plan was therefore prepared to bring new life into it and provide a contemporary environment to suit the firm's needs now and into the future. The original building had been Grade B listed by Historic Scotland in October 2005. This introduced significant challenges with

Figure 15.49 The refurbished building
Source: Arup

Figure 15.50 The original building before refurbishment
Source: Arup

regard to achieving the desired BREEAM 'Excellent' rating, and the re-development therefore became a detailed balance between preserving the qualities of Scotstoun House and meeting the needs of a twenty-first-century office building. This involved maximising the original building as a working space and complementing this with a new extension containing support functions. The link between old and new created a useful intermediate zone for break-out, group working and informal interaction.

The main principles of the development lay in opening up the original space, converting the open courtyard which it formerly enclosed for more offices and extending along the eastern boundary of the walled garden that lies immediately to the north. This east-facing new build component houses a new main entrance and reception area, meeting rooms, toilets, print room and staff area. The original stable blocks to the north beyond the walled garden were converted into mechanical and electrical plant space, facilities for staff showers, an independent remote conference room and a new cycle shed to store up to twenty bicycles.

The original building was constructed using timber joists, boarding and roofing felt, all supported on steel universal beams spanning between precast concrete wall units. The roof overhung the external walls to provide a cut-off from solar glare. The northern half of the new extension utilised the original east-facing stone garden wall, which required underpinning for stability and support. New precast concrete columns support 'glulam' beams connected back onto it. In the southern part of the new extension, structural steel links onto the original overhanging steel beams give a 1 metre band of glazing between the new extension and the existing structure. The overhead glazing allows natural light into the informal seating area below.

The team took a holistic approach to the design process, integrating the disciplines of architecture, building services and structural engineering. The most significant challenge for the project followed from the Grade B listing, which imposed several constraints such as retaining the existing wall and elements of the internal furnishings. In addition, the project was subject to the 2008 Scottish Technical Standards for energy usage, the performance requirements of which were difficult to achieve given the listing constraints.

The refurbishment contract was let in February 2009 and the main contract and fit-out period was seventy-four weeks. This construction period fell in line with the overall programme and the building reopened in July 2010. The staff had been decanted to temporary office accommodation within the grounds of Scotstoun estate.

Energy refurbishment

The new design was required to show innovation, creativity and technical excellence. The project objective was to create a world-class environment for employees and visitors, combined with an energy-efficient, sustainable design. To achieve this, the project team took a hierarchical approach, first optimising the building form, considering the building massing, orientation and envelope performance, thus enabling a highly energy-efficient building. Only once the building performance was optimised were the specific low-energy services systems selected.

Ventilation

Dynamic thermal analysis was used to design the building as fully naturally ventilated. The original office was 24 metres deep and great care had to be taken to control air movement through the redeveloped building. This was achieved by constructing in place of the courtyard, an atrium pod which draws air from the perimeter through the offices in a controlled manner. The natural ventilation strategy was also tested using bulk airflow calculations for various wind directions to ensure that the passive approach would operate correctly throughout the year. Solar-powered 'windcatchers' have been provided for the natural ventilation of internal meeting rooms. Night cooling and ventilation are used as and when high temperatures are detected during the night hours. The natural cooling scheme is further enhanced by the use of phase-change thermal mass incorporated into the ceilings as part of the insulation as described below.

Insulation

Even with the restrictions imposed by the listing, the U-values of the building elements are significantly better than required by the UK Building

Original Building

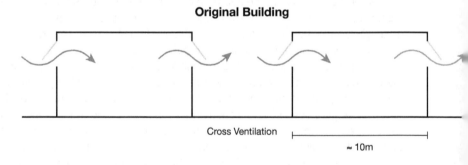

Cross Ventilation

$\approx$ 10m

Redeveloped Building

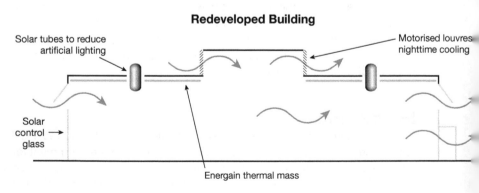

Solar tubes to reduce artificial lighting

Motorised louvres nighttime cooling

Solar control glass

Energain thermal mass

Figure 15.51 The new ventilation strategy

Regulations. A new format of insulation was installed above the main office ceiling; this material is a panel, composed of paraffin PCM, encapsulated in a unique DuPont polymer and laminated with aluminium. This innovative panel for the construction market is only 5 mm thick and was designed to be an effective and easy-to-install solution to the problems of excessive energy consumption and unpleasant temperature peaks in low-inertia buildings. DuPont™ Energain® thermal mass systems claim to save on average 30 per cent of air-conditioning costs, up to 15 per cent of heating costs and reduce temperature peaks by up to 7°C.

The windows were required to be reinstated like for like as the building is grade B listed; the windows originally were single-glazed and these were replaced with triple-glazed units, giving a U-value of 1.6 W/m²K.

Use of natural light

Rising energy costs, coupled with the requirement to reduce carbon footprint, make the need for feasible and effective ways to introduce natural light into buildings greater than ever. The deep-plan existing offices at Scotstoun House presented considerable challenges with regard to maximising natural

Figure 15.52 The sun pipes on the roof

Source: Arup

light, and this was remedied in the redevelopment by incorporating solar tubes or 'sun pipes' throughout. These introduce natural light at ceiling level, and give a 60 per cent reduction in energy consumption, as well as assist the natural ventilation strategy by reducing heat gains. The ninety solar tubes were specifically designed for the project and form a zero-energy light source of particular use in winter months when there is less daylight from the external glazing.

Artificial lighting

Supplementing the solar tubes throughout the main office area are direct suspended T5 luminaires (low-energy 5 mm diameter fluorescent lamps) with Cat-2 anti-glare louvres. Lighting control is by the integrated mechanical and electrical control system. The KNX switch regulates the DALI (digital addressable lighting interface) control gear within each fitting and interfaces with the daylight and PIR (passive infrared) sensors. When the natural light via the solar tubes achieves an illumination level of 350 lux, the control system regulates the artificial lighting. From dusk onwards the luminaires are activated, supplying a uniform level of 400 lux. Calculations indicate a resulting 60 per cent energy saving. In the open-plan courtyard space, recessed modular light fittings with single 42 W compact fluorescent lamps were installed with light sensor and movement PIR controls. Downlighters have been installed in the front of house area, again with PIR control. In the car park, low-energy pole-mounted LED fittings have been installed, with solar-powered LED road studs in the entrance road to the office.

Renewable energy

Planning permission required that a minimum of 15 per cent reduction in carbon emissions by the systems serving the building be provided by low-to-zero carbon (LZC) technology. As Grade B listing limited significantly the type of LZC system that could be used, this planning condition was achieved through the use of a biomass boiler.

Energy metering, monitoring and targeting

Comprehensive integrated control was required, and the chosen system successfully integrates the control of the electrical and mechanical services with comprehensive monitoring and metering. As an example, the gas and water meters are both monitored, the latter enabling any leaks in the buried distribution mains to be identified. The lighting system's electricity consumption is also extensively metered and the data from this help to optimise the lighting control, which in turn assists in maximising the performance of the daylighting strategy. The internal and external environments are comprehensively monitored for temperature, relative humidity, CO_2 and light levels. The system includes a weather station, the output of which has been used to validate the original modelling of daylighting against measured values.

Figure 15.53 A biomass boiler provides most of the heating

Source: Arup

Other environmental refurbishment

The existing building's structure and facade were retained and reused. As well as the length of garden wall incorporated as part of the facade of the new extension, another section of it was demolished, stored and rebuilt within the building as part of the redevelopment. On-site material was stored and crushed and used for granular fill and hardcore across the site and under road bases.

The Scotstoun House redevelopment maximised the use of natural, inert and recyclable materials, as well as elements of prefabrication to reduce construction waste. The reuse of the existing building, structure and facades, and the use of on-site crushed aggregates, minimised the need for new construction materials. The project made extensive use of Green Guide3 A-rated materials and insulation materials with zero ODP (ozone depletion potential) and GWP (global warming potential) of less than five.

Energy reduction and environmental achievements

The redeveloped building has achieved an EPC (Energy Performance Certificate) A-rating, a significant achievement for the refurbishment of a 1960s Grade B listed building.

The extensive use of the solar tubes throughout has led to a much improved working environment with a very uniform level of daylighting and less use of artificial lighting. The energy performance achieves a lighting load of approximately 10.2 kW over an area of 1,700 m^2, or 6 W/m^2. The lighting performance gives a task luminance of 400 lux, uniformity of 0.8 and glare factor of 19.5 UGR (unified glare rating).

The planning constraints required that the building achieve BREEAM 'Very Good' rating, and this goal was exceeded by the project being certified as BREEAM 'Excellent'.

Financial aspects

The construction cost of the project was £3.5 million, within the set budget. The financial total for energy savings has not been monitored, however the estimate is a saving of 60 per cent of the electrical energy.

The building in operation

The offices are maintained by a facility management contractor who monitors the energy consumption on a monthly basis. A post-occupancy assessment was carried out which gave a clear indication that the refurbished building gave the staff a far better and more environmentally friendly office in which to work.

The refurbishment is seen as a great success both by the occupants, the client and the outside world, proven by awards received from RIBA and BOC (British Council for Offices) for the best refurbished office under an area of 2,000 m², in Scotland and the UK.

Figure 15.54 The new office accommodation

Source: Arup

Figure 15.55 The new entrance

Source: Arup

REFERENCES

Baker, N. (2009) *The Handbook of Sustainable Refurbishment: Non-Domestic Buildings*, London: Earthscan.

ECON 19 (2003) *Energy Consumption Guide 19*, Energy Consumption in Offices. Available from: www.actionenergy.org.uk (accessed 27 May 2014).

INDEX

absorption cooling, 135, 237, 286, 289, 293, 295
adaptation, 3, 36–39, 46–48, 180, 205, 255
AHUs, 63, 280, 282
air conditioning, 75–76, 82–84, 86, 100–103, 106–107, 114, 118, 123, 137, 143, 150, 164, 197, 252–253, 307
air infiltration, 108, 114, 171, 175, 202
airtightness, 44, 49, 51, 156, 236, 259, 273, 277, 289
AITIM, 283
Apache, 178
appliances, 1, 55, 76, 78, 89, 192, 194, 197–198
argon fill, 131, 259
asbestos, 7, 61, 213–214, 223–224, 248
assets, 12–13, 16, 26, 28, 31, 34, 49, 61, 65–66, 68, 73, 85, 94, 147, 226, 269
atria, 5, 61, 65–66, 71, 104, 106, 128–129, 131, 151, 262, 265, 271, 288, 306

balconies, 69, 104, 131
BASS, 276
BCO, 17, 59, 270
behaviour, 20, 32, 38–39, 42, 49–51, 55, 101, 122–123, 191, 226, 252
BER, 212, 226
bicycles, see also cycles, 11, 268, 305
BIM, 189–191
biodiversity, 94–95, 160, 192–193, 199, 203, 212, 269, 280, 284
biomass, 131, 138, 183, 237, 262, 267–269, 286, 290, 303, 308–309
blinds, 5, 48, 104, 106, 113, 127–128, 134, 151–152, 242, 248–249, 302
BMS, 20, 133, 138, 244, 259–261, 277, 280, 283
boilers, 63, 86, 131, 138, 237, 244, 262, 268, 286, 289–290
boreholes, 55, 132, 252–254
borescope, 171
borrowed light, 128
BREEAM, 4, 6, 12, 68–69, 92, 96, 148, 156, 161, 163–166, 168, 206–208, 212, 216–217, 220–223, 239, 245, 256, 260, 262, 265, 269, 279, 303, 305, 310
BSRIA, 8, 223, 225, 229, 232, 235

Builder, see DesignBuilder, 177, 190
buried pipes, 132, 308
BUS, 240, 248, 268

CAD, 171–172
calibration of models, 173–175, 182, 185, 189
carbon emissions, 14–15, 17–18, 20, 27–28, 36, 43, 47–48, 59, 66, 79, 89–90, 93–95, 115, 140–141, 144–146, 148, 151, 157, 159–160, 170, 173, 184, 191–193, 198, 205, 216, 234, 247–249, 261, 267, 269, 307–308
CASBEE, 168
cavity insulation, 51, 130, 152–153, 157, 160
CCS, 220–221, 223
CDM, 213, 216
certification, 82–83, 165–169, 206, 220, 234, 283
CFD, 6, 181–182, 188
chilled beams, 50, 67, 119, 136
chillers, 63, 131–132, 135, 237, 267, 286, 289, 295
CHP, 86, 278, 280–281, 293, 295
CIBSE, 40, 42, 173–174, 183, 190–191, 199, 208
cladding, 51, 62, 131, 140–144, 152, 155, 245, 265, 267, 274
climate issues, 1, 3, 29, 34–43, 45–49, 51–53, 55–56, 72, 75, 87, 90, 94, 101, 103, 106, 108, 125, 133, 150–151, 169, 171, 180, 188, 191–192, 199, 205, 269, 273, 275, 280, 293, 302
clo, 180
coatings, low-e, 109, 131, 152, 259, 283
comfort, 2–7, 17, 20, 38–39, 42–44, 47, 50, 100–112, 115–123, 125–126, 133, 136, 140, 150, 152, 169–173, 175–177, 179–181, 183, 185–187, 189–192, 237–238, 248, 250–251, 253–256, 266, 271, 292–293, 296–297, 302–303
commissioning, 7, 43, 99, 163, 183, 210, 213, 222–223, 227–229, 232, 246, 248, 260–261, 277
condensation, 119, 130–131, 136, 142, 148, 155, 198, 207, 280
condensing boilers, 131, 244, 300